AF505859

Kissing Cousins

Kissing Cousins

A NEW KINSHIP BESTIARY

FRANCES BARTKOWSKI

COLUMBIA UNIVERSITY PRESS ♦ NEW YORK

Columbia University Press
Publishers Since 1893
New York   Chichester, West Sussex

Library of Congress Cataloging-in-Publication Data
Bartkowski, Frances, 1948–
Kissing cousins : a new kinship bestiary /
Frances Bartkowski.
p. cm.
Includes bibliographical references (p.   ) and index.
ISBN 978-0-231-14452-0 (cloth : alk. paper)
ISBN 978-0-231-51763-8 (electronic)
1. Kinship. 2. Familial behavior in animals.
3. Human–animal relationships.
4. Human–animal relationships in literature.
5. Human–animal relationships in motion pictures.
I. Title.
GN487.B38   2008
306.83—dc22   2008005266

Columbia University Press books are printed on permanent and
durable acid-free paper.
Printed in the United States of America
c 10 9 8 7 6 5 4 3 2 1

Frontispiece: Joan Fontcuberta, *Haemogram TG 15-5-98*

For Jules and his kin

CONTENTS

ACKNOWLEDGMENTS

I am immensely grateful to so many who have helped this book to see the light. Let me try to gather as many as I can. Since this is a book about kinship, all these individuals are part of a wide web of what is usually meant by *family*, and what for me constitutes connection of the most significant kind—under the sign of friendship. They have offered support, encouragement, criticism and insight where I live and work, where I have traveled and lectured, and in the classrooms where I have taught.

At home there is my son, Jules, who, even before he was born, began to transform my thinking regarding kinship in the most unforeseen ways; there are also the extended aunts, uncles, and cousins whose place in my life has led me to consider how else, besides the origins fate decrees, we choose to gather around us those who will enrich our being in the time we have to work and love.

Among the friends who have never wavered in their belief are Renae Bredin, Judith Butler, Lee Quinby, Jim Creech, Charna Meyers, Roy Nathanson, Harryet Ehrlich, Nancy Gerber, Susan Ghirardelli, Phyllis Greenspan, Alice Klein, Kate Stearns, Madeline Tiger, E. G. Crichton, David

Greene, John de Stefano, Sonja Wagner, Wendy Deutelbaum, Dee Morris, Donna Bassin, and Jessica Benjamin.

Among colleagues I want to thank Steven Diner for leading me to *Leon, the Pig Farmer*. Tim Raphael gave generously his "kintimacy" in reading closely several of these chapters along the way. I also want to thank Charles Russell and Virginia Tiger, my chairs during this time, for their support; the Program in Women's Studies at Rutgers–Newark has been a home to my work and my ideas, and Jyl Josephson has on more than one occasion offered the space to try these ideas out with faculty, students, and others. Also at Rutgers, Nancy Diaz, Barbara Foley, Kathe Callahan, Kimberly Da Costa Holton, Gary Roth, and Beryl Satter have offered friendship and ideas and suggestions that have moved this work along in crucial ways. An early version of the title essay was the focus of a seminar at the Institute for Research on Women at Rutgers–New Brunswick; that interdisciplinary group of feminist scholars and activists cheered me on in the early stages of organizing this book. Spencer McGrath began the bibliographic documentation for this book, and Melissa Metcalf took up and completed the work; I was fortunate to find both of these graduate students and happy to work with them, even under the dual constraints of time and remuneration for their expertise.

The earliest version of this work germinated in the rich feminist universe at the University of California in Santa Cruz, where so many were actively involved in the remaking of kinship and community, as well as in theorizing questions of conflict and connection. Later, invitations to lecture on this work in progress were instrumental in its development. An impromptu invitation to speak about bonobos at the Center for Twentieth Century Studies in 1997 led to my spending precious time with the bonobos and their curators and keepers at the Milwaukee Zoo, which then housed the largest group in captivity in the United States. Miriam Silverberg organized a conference on Gender and Intimacy in 2003 at UCLA, where I took heart in comments offered by anthropologists and historians. In 2002 David Shumway and former colleagues at Carnegie Mellon University were good enough to invite me to talk about the focus on primatology that had begun to emerge in this work. The conference "Kinship in Europe: The Long View," held in Ascona, Italy, in 2002, offered a heavenly venue and an amazing group of scholars from numerous countries from whom I learned more than I can say about how to continue my work on questions of kinship. In Toronto, at an international feminist conference in 2000, I came to realize how this new work was intimately tied to my earliest scholarship. In Edmonton, as a guest of the English department, I was warmly received and inspired to continue looking for stories to add to my growing bestiary.

At the "Kindred Spirits" conference in September 2006, Donna Haraway was kind enough to suggest a trajectory for this book, which led me to a brief but critical encounter with Cary Wolfe at Rice University and ultimately to Wendy Lochner (and her colleagues) at Columbia University Press, whose belief and blessing have been invaluable; the readers she chose were veritable angels in their willingness to offer pointed remarks about what would substantially improve this book, and their praise for what I had done helped me immeasurably in going back through it again to strengthen as best I could all that has propelled my energies over the past decade that this book has been in the writing.

Numerous others ushered and cheered me along in my curiosity and commitment to seeing this project through to its end.

Acknowledgments also to the *New Yorker* Cartoon Bank, Getty Images, Frans de Waal, Harvard University Press, Patricia Piccinini, and the photographers Graham Baring and Joan Fontcuberta.

Kissing Cousins

The total humanization of the animal coincides with the total animalization of man.

—Giorgio Agamben, *The Open: Man and Animal*

More than forty-five years ago, Carl Reiner and Mel Brooks invited people into their studio, where they improvised on several occasions this imaginary visitor from another time. In the course of these "interviews" with a "man who claims to be 2,000 years old," Reiner asks Brooks to elaborate on what life was like back then, when folks lived in caves. Pursuing several lines of questioning, he also inquires whether the people who lived in these caves had ways of celebrating their existence: Did they have national anthems, for example? Brooks, in his Jewish-accented man of the past replies, "Of course, we had national anthems!" feigning astonishment that anyone would doubt it. When Reiner asks whether he remembers his national anthem all these years later, the old man, even more pseudo-shocked, says belligerently, "You don't forget a national anthem, dummy!" Reiner then asks if he'll sing it for the audience of listeners rapt with interest in what this man brings back from his long past. He is glad to oblige and belts out a one-line song, the tune of which you can perhaps imagine from the rhetorical strategy it uses. He intones, with mock gravity and pride (while we

"You are here." (Cartoon by Robert Leighton, *New Yorker*, December 25, 2006)

can hear the two comics holding back their own laughter), "Let 'em all go to hell, except Cave 73!"

I've been remembering the 1960 visits of the 2,000-year-old man. And though I'm not one for telling jokes, this joke keeps me laughing even as it has haunted me these past few years as I have been writing about what kinship looks like in these early years of the new century. If I can sum up the purpose and perspective of this book in a joke, I wonder, then, whether the book itself becomes superfluous. But perhaps the book is the inevitably less humorous endeavor to explain the joke to those who don't get it or to those who might want to contemplate some of its premises. For I think they concern matters of life and death.

Where are we in our understanding of the place of the cave in culture? There is an arc from our early hominid history to its first philosophical appearance in Plato to its bridge to the postmillennial present. Caves have no doors. More solid than nests, more vulnerable than houses, they shelter *and* hide. Does the 2,000-year-old man sing his anthem to those inside, his back turned to those outside? Or is he more likely guarding the opening against those who might approach? The anthem is the voice of the collective, the nation, the state, rendered musical. Those in the dark voice their presence to the ones outside. Or the one standing guard outside intones his song to reassure the ones cowering, sleeping, feeding, mating, painting in the dark. The cave is the enclosure, and those outside its walls may be worth killing so that those inside may live. And this is a zero-sum game.

Isn't this a joke we all need to be in on so we'll be living on, rather than killing off one another? At a time when walls and fences and barriers are being erected in bricks and mortar and electronic circuitry, we might rather ask what the world would look like if we opened the borders of our selves to the others—those others whom we resemble though we may be inclined to insist that we do not recognize them as our coevals, our co-evils. Our cousin-citizens. Our "concestors," as Richard Dawkins calls them.

Within the same world-historical week in London in July 2005 two utterly contrasting things happened. Both were meant to change the world, and both were addressed to those gathered at Gleneagles, Scotland, for the G8 meetings. Consider that one method summoned tens of thousands into the light (in London and around the world), making music in their effort to get "eight men in a room" to consider carefully their power to change the daily conditions of millions. Later in that same week, we want not to forget, others worked in darkness and anonymity, touching off sounds of alarm and the keening of countless others through fear.

Our times urgently require that we think our way out, outside of and beyond the cave. We need to think kinship, intimacy, "kintimacy" from the primal to the millennial, from the prelapsarian to the not-yet, a category founded on hope. The "circle of empathy," theorized by Jaron Lanier, ought to be ever widening, all the while we know it to be a circle that includes both conflict and connection. Let me suggest the 2,000-year-old man as a holy fool, a jokester uncle whom we invite for a visit. To leave his cave. To listen, to laugh, to grieve, to ponder, to imagine a world of "kissing cousins."

PART I

There are gestures that decide people's lives: the handgrasp, the kiss . . .

—Nadine Gordimer, The Pickup

In genealogy you might say that interest lies in the eye of the gene holder. . . . Yet everyone loves stories, and that is one reason we seek knowledge of our own blood kin. . . . We have climbed back so far into our family trees, we stand inches away from the roots where the primates dominate.

—Ellen Meloy, The Anthropology of Turquoise

1   KISSING COUSINS

To talk about kinship is to invest in the future of our lives together. And the stakes in that investment ought to be legible. So let me say a few things about my personal and professional stake in these matters.

In its earliest form this project was born out of my encounter with the practices of what were then called "new reproductive technologies." Though I was availing myself of some of the lowest "tech" versions of these procedures—anonymous donor insemination—I could not help but be provoked to face some of my own deeply held and unexamined assumptions, as well as the assumptions of those surrounding me, as I ventured to become a "single mother by choice," a phrase of the 1980s that sounds quaint to my ears today. By the mid-1990s, approaching the "age of reason," my son could already offer an explanation on the playground if anyone challenged his statement that he had "no dad." Of course, I, his mother, had had some years to think her way toward constructing such a family romance, with its seemingly "missing" piece. I had first approached this challenge most concretely on a day in 1987 when I finally got down to filling out the required form (never a task I warm to) for that anonymous—but known to some technician in New York—donor who, if I was fertile and healthy, would become my child's "donor," not "father."

As a well-trained feminist, I had learned in the 1970s to separate sex from love; now I was schooling myself in separating the biological from the social, and not in theory, but in the facts of what would become my life as a mother. So I knew my family would not resemble that vanishing species of the married heterosexual reproductive bonded pair, but I also knew in my devotion to relations of intimacy that my son and I would be making kinship among whomever we might live. Nearly twenty years later, these questions have receded far into the background of what happens to be simply our kinship story. But in one form or another, the attention I gave to such issues then has led me to follow closely the many ways that knowledge, science, and changing social forms have continued to produce relations of family, intimacy, and kinship that dare us to confront some of our most preciously held ideas. Many of these ideas remain central to debates and struggles we are watching played out daily still, and daily and still they require us to become ever more literate in our notions of who may be related to whom.

This book is written in praise of mixing, for we are, I think, misguided to engage discourses of purity in an age of chimeric realities. We live in a time of donation, surrogacy, adoption, genetic transfer, splicing, and design. Practices emerge and become conventionalized while ethical debates and disturbances and regulatory statutes and strictures follow behind. State interventions are subject to political agendas that change with electoral seasons. Some nations lead the way, some commercial interests prevail over policy discussions, and some individuals decide they know how we should address the futures suggested by our present ways of knowing.

As a student, teacher, and critic of literature, I am interested in and curious about the stories we tell and how we tell them. I am perennially caught up by the stories we have been learning to tell, retell, and untell about all our kin. Our kind. We women. We humans. Those we call our next of kin. Those we claim through ties of blood, flesh, bone—genes, chromosomes, peptides. I think we might all agree that we are living through a transition of relatedness, connection, proximity, permeability, and resemblances. Where some delight in the confusion of boundaries, some patrol the borders. Kinship is a category intensely critiqued yet remade for and by feminist inquiry. Among feminist anthropologists the category has been thoroughly rewritten and reexamined by virtue of two contemporaneous strands: the internal critique of kinship study as a hegemonic Euro-American imposition of codes of blood, and the simultaneous interrogation of how reproductive technologies have rescripted the heteronormative domestic sphere of assisted reproduction. My focus is on the borders and boundaries that emerge from a wish to remap our connectedness, our relatedness, subsequent to our fevered studies of differences, of divergences. The identity politics of the late twentieth century focused our attention at the points of difference even while demanding the coalitional, collective work across those differences; the early twenty-first century has us turning from otherness and difference to relationality and sameness, back some distance from the particular, demanding the glance that can take in universality, all the while refusing totality.

Kinship touches deeply on bodily states: the phenomenological, the sensory, the questions of boundaries—at the skin, under the skin, blood thicker than water, the brain, the bodymind, the natural made cultural, or "cultural naturals," as named by primatologist Frans de Waal in *The Ape and the Sushi Master*. What would it mean if kinship were repatterned over designs we might recognize under the sign of friendship, for example? Questions of kinship raise the stakes on how cultures always and everywhere choose to make flourish that which is deemed good in ourselves and others. Among kin is where we are called on to be our best. It is also the sphere where humans fail to answer the call as friend and where, instead, kin may be transformed into foe—the outsider, the intruder.

The essential interdisciplinarity of this project is everywhere visible in its willful permeability and vulnerability as it borrows from philosophy, for kinship is a study of the ethical, that which is proximate and worth caring for and struggling over; anthropology, for kinship cannot be derived without serious consideration of the arrangements of intimacy that are the manifestations and symptoms of its structures; politics, for kinship presides over relations to those most vulnerable—the young, the aged, the ill, those who require care of others—who insist on our interdependency; economics, because kinship is inextricably tied to exchange systems and how they prevail on who owes what to whom in the arenas of labor, production, circulation, distribution of the goods; and, of course, the biological sciences—what Donna Haraway has suggested are the new humanities—for kinship is where we wade into the deep waters of what we make of that which is given. Driven by curiosity, the upside of ignorance, I have learned particularly from three disciplines not my own—anthropology, genetics, and primatology—so that I could return to what I know well: narratives, stories. This book, then, is a kind of accumulation of stories found where I have been drawn to look and listen.

"Kissing cousins" occupy a liminal space in our notions of kin, suggesting boundaries between those we may kiss and those we ought not to; *Webster's Third* frames it thus: "a cousin or collateral relative whom one knows *just* well enough to kiss *more or less* formally upon occasional meeting" (emphasis mine). The question is *who may be related to whom*. We need and find new names for our intimates, as the norms shift and the majority of families no longer resemble the two-parent, two-generational, so-called nuclear family.

Transgression and the uses of technologies high and low have produced new stories of intimate connections. Connections across species boundaries are currently transforming the stories we tell to help us speak these interrelations. Global movements have produced new kinship networks that defy easy categorization. The central dynamic is between what we consider *natural* and what we assume to be *cultural*. Both these categories exemplify our human need for boundaries. At the moment all such sites are under reconstruction, renovation, and reinvention. Among the questions raised by these changes are (1) how are stories of origins being retold in times of no longer new reproductive technologies? and (2) how are reinvented "totems and taboos" helping us deal with categories that have been destabilized? Narratives that emerge from our ever-shifting scientific knowledge come to bear on how we live in the brave new worlds that are not fictions. In *Kissing Cousins* the sources for these exemplary narratives will come primarily from three arenas of public culture: the news, film, and literature.

In recent decades families and kinship relations have shape-shifted as dramatically through reproductive technologies and choices as they did in the period of industrial revolution. The nearly infinite variability of forms for our closest relations is the object of my study here—and in the most intimate sphere—those we come to love and care for through association driven by need, desire, and demand. Again, as a critic of literature, I understand that my intervention is at the level of story, storytelling, and its permutations of kinship narratives. What do they resemble today? The contested sites remain, and their contestations attest to the ferociousness with which we cling to the kinship narratives we claim. What is it to be brothers? Sisters? Cousins? What are we naming in this lexicon? We want to know the countless ways that others before us have reckoned (with) proximity. How do people make a language, a system of meanings out of who they are, who they've become, who they might have been, and what lines of affinity and affiliation have been written and codified? It has been an article of faith in feminist thought that kinship is the preoccupation of patriarchy, fathers and sons, so as to be able to claim, stake, name, and thereby render a matter of law, the name of the father. But it would seem that history has outrun some of the thinking about family and kin and its cultural work, because if it is a crisis in kinship we are witnessing, and I would submit that it is, then it is a crisis precipitated to a great degree by the unforeseen destabilization of maternity and motherhood that we are witnessing in the early years of the twenty-first century. Consider surrogacy and egg donation as the salient examples that rewrite the realm where we thought we knew with a kind of certainty what was indivisible—the one whose body mediated between the self and the law.

The anthropologists Sarah Franklin and Marilyn Strathern are among those who have taken up David Schneider's critique of folk-inflected notions that emerge from "the facts of life": the biological and its naivete, and hegemony. Numerous anthropologists, in returning to kinship studies, have taken up the critique of the heteronormativity at the heart of earlier studies, such as Schneider's. Strathern's work in Melanesia opens onto questions of how the self is conceived in a culture where, as she notes, the person is not seen as an individual—our fetishized Euro-American notion of personhood—but rather as "dividual." What would it mean for kinship if the reproduction of selves were based in another economy of exchange and movement of selves among members of a community? The cultures of science, from biologists to primatologists, engage dialogues across boundaries that we can no longer presume to be impermeable. In *Antigone's Claim*, Judith Butler recalls to our attention the fact that kinship matters are matters of "life and death." I aim here to trouble the waters of kinship based in narrative economies of blood. Doing so will require exploring some of the material and conceptual effects of the successes of technological replies to demands for reproductive choice and freedom. Notwithstanding the destabilization of the category of maternity at this century's turn, what remains intact is that the uterus remains a site where the body and the law intersect.

In the year 2000 there was an ongoing narrative on the adjudication of an in-vitro fertilization error that had produced kinship repercussions in two separate families. In the lawsuit that followed, visitation rights were denied to a white woman who gave birth to a black woman's baby after the black woman's embryo was mistakenly implanted in the white woman's womb. Is this surrogacy by accident? Is this miscegenation, with all that suggests about the ways we calculate race, motherhood, and tenancy in the womb? What does it mean to say that the courts and psychologists will decide "whether they [the twins?] had formed links as brothers during 14 months in the womb and the . . . home." If the womb is the first home, and the home is suggestive of maternal space with all that implies about our deep attachments to stories of origins, what meaning do we make of the split between social, biological, and genetic mothers? How do we understand the spectrum from egg donors to surrogate, lesbian, adoptive, birth, and foster mothers? How does the distinction between bio- and non-biomoms come to have meaning? What would we feminists have these distinctions mean in the ongoing ethical questions raised by the last two decades of "advances" in the technology that gives and errs? We are beginning to hear about the practice of unwanted/disused embryos being adopted by those whose antichoice views on the "right to life" have been organized by projects such as Snowflakes and Embryos Alive—another approach to the consequences of technologies used by those seeking fertility. For some women in the world, access to the technological has meant abortion rather than infanticide. For others, men and women, gay and straight, this has meant children to raise who were available through international adoption and contract surrogacy. These practices are also the outcomes of global politics, markets, and migrational patterns—daughter disfavor takes different forms but remains a defining feature of cultures of deprivation. Where lives are at stake, female children are inevitably more at risk, and subject to traffic and rescue. How do we negotiate these moves from personal desires to free markets to courts of law?

If this narrative verges on the solomonic-theological, here is one that touches on the borders between species that also concerns us here. Again, from the nonfictions of everyday life, on July 19, 2000, the *New York Times* reported that the Flikshtein family of Brooklyn, New York, went to court to prevent the removal from their home of Cookie, their adopted primate daughter who had been living with them and was beloved by friends and neighbors. That she is a member of an endangered species only made her protection more necessary. Here the forces at war were the family: "They want to take the baby from the mother. But she will die without us, and we will die if we give her up." The family was in opposition to the experts, who said that "she does not belong in human society." What are the forces at work that tell us of our humanisms, our humanity, even as they reveal the inhuman, subhuman, and superhuman that we are as well? Stories such as these demonstrate just how deeply etched our interconnections are. What will be the currency of kinship in this millennium? While it may be that kinship is "a precondition of the human" (Butler, *Antigone's Claim*, 82), and systems of classification of those who are closest to us appear to be "ubiquitous cultural product[s]" (de Waal, *The Ape and the Sushi Master*, 285), how are we witnessing and claiming the space where matters of kinship are determined, adjudicated, and always redefined by need, choice, desire?

Exceptions rewrite the rules about desire. How might we gather ideas about how the world could look anew? What forces would constitute our relations across and through differences? These are the questions I was asking twenty years ago in my work on feminist utopias. The idea I was tracking was to see how a group of women writers, in the worlds they imagined, gave shape to social life. More explicitly, I was asking how, if the family functioned as a microcosm of the state, did these writers reshape, redesign, rearrange that nexus of interrelations known as family? What made intimates of strangers? What were the bonds between old and young? How did imaginary citizens construct (social and political) life in the "good place," as it was defined in the period under consideration—largely the 1970s? Donna Haraway produced what many consider to be the most pathbreaking piece of theory of the mid-1980s in "A Cyborg Manifesto." Even now, in the year 2008, her voice within feminist theory is unique in envisioning the world as changed by a new consciousness—one informed by networks of new technology, new power relations, and replaceable parts that facilitate the remaking of bodies in space and time: "The cyborg story raises questions about our kin among the machines—our kin within the domain of communication—while the primate story raises questions about our kin in the domain of other organisms," she says (Haraway and Goodeve, *How Like a Leaf*, 105–6), once again complicating the questions of kinship as they impinge not merely on the "facts of life" but also on the facts of technologies that we have made and that remake us. We must learn to exercise an understanding of some of the signs by which we can read the dialogue between past and future here in our present.

Technology and kinship both might be deemed what we make of that which we are given—the potential derived from the possible. Isn't it the case that the daily news presents us with phenomena that surround us but that we don't necessarily comprehend, yet about which we all are able to offer opinions? And isn't it a kind of continuing education in emotional and social literacy to be in a position to look around and see what is happening in the realms of kinship and to read and interpret those stories? For questions of utopias and dystopias haunt this century, as they did the last.

To circle back to my stakes and sources in this project: it was some three decades ago that I had been tracing avidly questions of the reconfigurations of intimacy that the feminist movement, theory, and inquiry had placed before us. Yet my twenty-first-century students are as shocked, if not more so, as were the 1976 readers of Marge Piercy's *Woman on the Edge of Time* when they follow Connie Ramos, Piercy's time-traveler, to a feminist utopia and find birthing chambers where embryos grow from fetuses into babies born ectogenetically—outside the body of a woman. Piercy imagined that "if women were to gain power they had to give up the one power they had always had—that of reproduction" (62). No, we are not quite in this kind of future three-plus decades later, but then the future, for all our imaginings of it, is precisely that, an imagining. In fact, it rolls out before us as the eternally present. With the publication of Margaret Atwood's *The Handmaid's Tale* (1986), it seemed that the dialectic of utopian thought had turned in cultural exhaustion and alarm to the dystopian. Atwood dressed her handmaids in red cloaks and covered them with outsized headgear so that they could barely move. So covered were they that they could speak only in coded whispers while fomenting resistance to the theocracy under which they lived. Feminists then and now pose questions that deal with the making of subjects/citizens for the future, for that has been among the labors of women. But we know that the variations on the divisions of labor and power are as many as the world's cultures; we also seem to know beyond a shadow of a doubt that this division is not made symmetrically anywhere among humans or among those species whom we acknowledge as our evolutionary coevals. For it seems the utopian has departed from fiction to the domains of the scientific. Where there was the malleable social, we now have the constantly morphing biotechnological sphere that asks the questions about how and what we might share across and through differences.

How does the examination of our distant pasts reshape our itineraries into not-so-distant futures? Evolutionary biologists (Robin Baker, *Sperm Wars: The Science of Sex*), zoologists (Richard Wrangham and Dale Peterson, *Demonic Males: Apes and the Origins of Human Violence*), and fiction writers (Jeffrey Eugenides, "Baster"; Simon Mawer, *Mendel's Dwarf*; Robin Cook, *Chromosome 6*), as well as primatologists (Roger Fouts, *Next of Kin: What Chimpanzees Have Taught Me About Who We Are*; Frans de Waal and Frans Lanting, *Bonobo: The Forgotten Ape*), have presented us with texts that are provocative in their implications for how we think about the cultural forms nature takes.

Kinship is the place where lines of filiation, consanguinity, and affinity come together. And incest is that site where law intervenes in these arrangements of intimacy. It is that to which any group/culture finally says, no, this shall not be; even among the bonobos, our closest primate kin, it would seem mothers and sons are a no-no. The nay-saying has the power to shame, dislocate, exile, and banish, as well as to indicate that which binds. In *The History of Sexuality*, Michel Foucault described how the regime of blood had spent itself in genealogy. Certainly the regime of sperm-reckoning shares the field with eggs and their anonymous, yet carefully selected, young donors. Now, with the human genome mapped, a universe opens in which we must acquire new forms of literacy to know how to read our way through the rapids of bioethical, biopolitical, and biosocial changes as swiftly as they emerge.

These current conditions form kinship networks that we must learn to understand. For blood, sperm, and genes are and will be mixed, spliced, and reconstituted in ways we are barely able to name. Changes in our very skins, trade in our organs, and newer prosthetic and surgical techniques will remake our very notions of what being alive and being human mean. Such practices make it necessary to learn to speak in new languages of those we come to know and love. And here I will explore some of the cautionary and curious tales from which we might learn.

2 FORGET THE ALAMO . . .

John Sayles's film *Lone Star* (1996) tells a tale that urges us to reexamine some dearly held assumptions about the relation between love and the law. About those we kiss and those we kill. The film is set in a Texas–Mexico border town named Frontera, where history is alive and well and highly contested, and one of the tasks Sayles reckons with is to teach all that must be remembered before it is possible and, yes, desirable to "forget the Alamo." A family, a town, and the borderlands of two nations must first know what blood has been shed and by whom and then decide what those losses and lessons will mean. What begins as a murder investigation ends at the question of love; we move from homicide (who has killed?) to incest (who will kiss?). Three chronicles of "blood relations" provide the threads from which this cinematic tapestry is woven. The richness of this American film text mixes epic narratives from North American history, family history, and film history.

Certainly the metaphor of blood and its shedding is key to the military history of the United States and Mexico, as much as its mixing is to the demographics of border lives.

Blood, its metaphysics and meaning, is also crucial to the Western, detective, and romance genres out of which Sayles stirs his tale of the founding of a civilization. And blood remains the most tangible bodily fluid through which many people reckon those with whom they are kin. Though currently being everywhere displaced by genetic mapping, it is crucial for my argument here that we remember before we go on to forget—remember that blood codes are among the first laws established in many cultures. With their universal yet culturally specific rules attached to what Freud named "totem and taboo," those congruent rules demand exogamy and forbid incest. You must "go outside" to engage a future—but not too far outside.

In the director's statement included in the press packet for the film, Sayles echoes the moves of the film:

> A lot of this film is about history and what we do with it. . . . You can get six different people to look at the Alamo and they have six different stories about what actually happened. . . . The same goes for your personal history. At what point do you say about your parents, "that was them, this is me." . . . That's also what this movie is about. In a personal sense, a border is where you draw a line and say this is where I end and somebody else begins.

Family has often been read as a microcosm of the state. Laws are instituted through connection and conflict, only later codified as law for the larger tribe, community, or nation. Freud's text on kinship tracks the significance of how kin relations presuppose "participation in a common substance," which not only is the blood and milk of one's mother but extends to food shared. Laws are rationales for the originating act of cultures: the murder and feasting over the body of the dead father—whose power has been successfully and necessarily usurped—by the band of brothers; that is the plot of *Totem and Taboo*. The making of families also represents our own nuclear diasporas; kinship scatters us while binding us, in conflict and through connection.

In the cinematic universe of *Lone Star* the local kinship network represents a unit(y) that becomes progressively more remarkable for its slowly disclosed dispersion among races and across the borders surrounding this small town. The reconciliation and mapping of these borders fuels a narrative addressing three generations of white, Chicano, and African American families. Simultaneously, we follow three generations of the local law, represented by a sequence of county sheriffs and overdetermined by the fact that the current holder of that elected office is the son of the one immediately preceding him: Sam Deeds (Chris Cooper), son of Buddy Deeds (Matthew McConaughey). Sam's performance of his state (county) function makes him the agent of a new historiography, a man who will rewrite the foundational narrative of taboo and its repression. Sam knows that his father, when young, had threatened the life of then-sheriff Charlie Wade (Kris Kristofferson) on the night Wade was killed. Sam has sworn to exhume the secrets of his own legacy, as son and as successor to the position of sheriff. It is noteworthy that on the Web site of the Texas Military Forces Museum we are told that the declaration of independence of the Lone Star state was "signed in the blood shed" at the battle of the

Alamo—which history and Hollywood have admonished us to "Remember!" We might ask the costs and benefits of a history that insists on, even makes it imperative, ideologically, to "remember!"

To make out what concerns me here, some circuitous moves through plot are necessary; I ask your attention and forbearance in following these tales of blood, food, and deeds. The film opens onto a "demilitarized" zone, once a rifle range for the military base in town, where two off-duty army officers are taking time out, one of them an amateur sculptor who melts used bullets, equipped with his metal detector; the other is a dedicated botanist, taxonomy text in hand, identifying obscure cacti and other plants native to the region. Their discovery of a rusted badge and a nearby desiccated corpse sets the narrative in motion. Three genealogies will become entwined before the knowledge of the deeds done is fully disclosed. One tracks through the men of the mixed Native American and African American Payne family, one through the Mexican and Chicana women of the Cruz family, and the other through the white men who have held the post of sheriff of Rio County. Legendary in Frontera, Charlie Wade, the first of the sheriffs in question, "disappeared" one night some forty years ago—the discovery of his rusted badge resurrects him. Key elders in the town still "dine out" on tales of Wade's murderousness and corruption. Beginning his informal investigation of these "remains," Sam knows where he can probably find some of the elders who remember the night Wade left town.

Sharing food, drink, and stories at Mercedes Cruz's (Miriam Colon) Santa Barbara Café are Hollis Pogue (Clifton

James), the current mayor, and his cronies telling their oft-told tales yet again. Caught in a guilty moment, they are, in fact, speaking with some misgivings of Buddy Deeds's literal legacy in the figure of his son, Sam, described by Hollis's pal, Fenton (Tony Frank), in the local idiom as "all hat and no cattle." Hollis tries to demur at Sam's request to tell the story once again of the night Charlie Wade disappeared. The first of several long flashbacks, this one begins with Hollis reminiscing in voice-over: "It all began over a basket of tortillas . . ."

The quest of the current sheriff, Sam Deeds, is to find out who murdered Charlie Wade, the man who held this office, was unanimously despised and feared, and preceded Sam's own father's tenure as sheriff of Rio County. Sam, the son of the much admired, even idealized, Buddy Deeds, is the third in this genealogy of Rio County sheriffs, each of whom represents not only a version of Anglo masculinity in post–World War II America but also a congealed political economy of governance. While the politics are local, they nevertheless echo national policies of the decades in which each man ruled. Charlie Wade was "one of your old-fashioned bribe-or-bullets kind of sheriffs," says Hollis Pogue, who, in his youth, had served as Wade's deputy. Wade, dead nearly forty years, was reviled by all but especially by those over whom he brutally exercised his power—the Mexicans and blacks in town. That basket of tortillas not only held food but hid the bribes Charlie and his deputy would come to collect on a regular basis from local small business owners.

Through his investigation Sam begins to suspect his own father, Buddy Deeds, as the murderer of Charlie Wade. To "get the full picture on Buddy Deeds," Sam must also pay a visit to the neighborhood in Frontera known as Darktown, where he searches out the original owner of the local African American drinking establishment, a bar now known as Big O's. In response to Sam's prying, unwelcome questions, Otis Payne (Ron Canada), the current owner, claims, "I don't recall that a prisoner ever died in your daddy's custody. I don't recall a man in this county—black, white, Mexican—who'd hesitate for a minute before they'd call on Buddy Deeds. . . . More than that I wouldn't care to say."

We come to know Otis through flashbacks to his youth as a defiant black man in the Jim Crow South that was Texas, now a man of gravity who occupies the unofficial position of "Mayor of Darktown." Otis has researched his family history and has built an informal museum of visual artifacts meant to trace (t)his particular mixing of African Americans and Native Americans. Otis also maintains a "shrine" of photos of his son, Delmar Payne (Joe Morton), who has recently come to town, with his wife and son, as the colonel in charge of closing down the local military base; this son grew up separated from his father, and their difficult reconciliation will be another of the kinship stories interwoven into the film. The youngest Payne is Chet (Eddie Robinson), a high-school student, who knows he has a grandfather in town and wonders why they haven't been introduced. In defiance of his father's military-style parenting, Chet wanders into Big O's bar on a weekend night that turns violent; his grandfather spirits him out before the police arrive. Later, Chet returns to explore the sign at the back of Big O's that advertises a "museum." In this pedagogical scene with his grandson, Otis explains that the Payne kinship network is made up of contemporary African Americans who long ago lived with, fought beside, and intermarried among the Seminoles—a typical borderlands mixing, forgotten until unearthed by Otis's amateur historical research.

What Otis won't yet, but finally will, say reveals the most pertinent secrets of blood shed in this local history—stories not fit for his "museum of mixing" or for the newspapers—rather, he bears the weight of the secret he pledged to keep as a vulnerable young man who has now become a leader of his community. When Sam comes to question him about a night years ago, Otis explains that Buddy's rule, which lasted for thirty years, was unlike Wade's violent tenure as sheriff. Buddy was interested in the "big picture—county political machine, chamber of commerce, zoning board." By this we are to understand that although Buddy's authority may have been exercised without threat of bodily harm, it was never in doubt: as long as Otis got the people of Darktown out to vote, Buddy Deeds "was pretty much on [Otis's] side. Whenever somebody thought of startin' up a new bar for black folks, they'd be, ah, how should I say this? . . . 'officially discouraged.'"

There is the kinship network of the Cruz women we must yet find out, before the full implications of Sam's quest can become plain. Mercedes Cruz—once an illegal immigrant herself but now an assimilationist—is the first to call the border police to report "wetbacks." Mercedes was widowed when her young husband was stopped on the road coming back to town from across the border by Charlie Wade and his young deputy, Hollis. Wade shot Eladio Cruz (Gilbert R. Cuellar Jr.) in cold blood. Pilar (Elizabeth

Peña) is Mercedes's daughter, and like her mother, Pilar is widowed, not by Charlie Wade but by the Vietnam War. We see some of the tension between mother and daughter when Pilar, the high-school history teacher, suggests a family trip across the border, because her teenage son is curious about his "roots." Mercedes finds no comfort in the thought of returning to the place she fled in another time, before ethnic pride made such journeys commonplace; to Pilar she hisses, "You want to see Mexicans, look around." Pilar reaches farther back into their past and reminds her mother that none of the men she ever dated as a young girl was good enough for this migrant/immigrant mother—not even Anglos. Mercedes replies, "It was just *that* boy." At this moment in the film it is still ambiguous who the boy in question was; it could have been "Nando," Pilar's husband, who died in Vietnam.

As the local high-school history teacher, Pilar struggles to do her best to present a "more complex picture" of local, state, and international history, which is precisely what displeases the Anglo parents, a distinct and anxious, but still powerful, minority. We watch their disputes at a school-board meeting over the "men who founded this country" as the complexity and completeness of versions of the past are recounted by the town and its citizens—adamant teachers and warring parents, observed by Danny, the ethnically proud Chicano newspaper reporter on the scene. In one particularly tense moment, a white mother allows as how "when it comes to food or music" she's unperturbed by the intermingling of her children with "others." This shift from the debates over history to the readiness to share the lunch table mirrors one of the leading metaphors of the film—the culinary aspects of totem and taboo. When Chet Payne's mother pays a visit to Pilar to see about the climate of the new school for her African American son, Pilar acknowledges that "blacks are the smallest minority, except for a few Kickapoo kids."

The romance that drives this epic history emerges when Sam runs into Pilar at the police station. Their looks are charged and warm; she is stressed, and he comes to her rescue, as he intervenes in a false arrest of Pilar's son. In this scene we are witness to the reigniting of a connection long broken yet remembered over time and distance. Twenty-three years earlier, Pilar and Sam had been high-school sweethearts, and they haven't spoken since high school. Sam is now divorced and childless, and Pilar is a widowed mother of two teenagers. They will meet next at the ribbon-cutting ceremony where a memorial has been dedicated to Sam's father, Buddy. Present for this occasion are all the local bigwigs, and Mercedes, now on the town council, cuts the ribbon. She watches as Sam and her daughter walk off together in the middle of this workday, and the look on the mother's face echoes the look in her earlier argument with her daughter. Sam and Pilar trade recollections of how strongly Buddy objected to their coupling as teens. Pilar says she was afraid of Sam's father; Sam expresses his regrets over his early years of defiance, now transformed by his wearing the same badge.

Sam confesses to Pilar that he doesn't really want to be sheriff, that what "folks wanted was the name, and someone to defeat the other guy," but now he feels that he is "just a jailer . . . [who] run[s] a sixty-room hotel with bars on the windows." The sheriff and his powers are up for grabs in the next local election; anticipation of change in this county office rests now with Sam's deputy, Ray Hernandez (Tony Plana), who understands that "our good time has come." By the requisite sharing of power with the white and Chicano businessmen and -women in town, he is likely to be the next and first Chicano sheriff of Rio County, beginning a new and reinflected fourth generation of the embodiment of law in Frontera. It may be tempting to critique this line of succession as the bribe of liberal humanism with its premise/promise of equality of opportunity and the granting of rights. Patricia Williams has noted that rights, while insufficient, nevertheless remain "islands of empowerment" (*The Alchemy of Race and Rights*, 233). The key word here is *islands*. Sayles's cinematic world of borders necessarily defies the image of islands as isolated and requiring exchange with some "mainland." In Frontera it is the metaphor of the stew, or *menudo*, of peoples that leads the way through change.

There is yet another elder whom Sayles privileges; he is the sole indigenous character featured in the film, Wesley Birdsong (Gordon Tootoosis). He is tenuously attached to the consumer culture, running a roadside stand of Native American paraphernalia, with a degree of irony. He tells Sam that the tourists won't buy much from him any more but that "in Santa Fe buffalo chips go as fast as they can squeeze them out." Birdsong knows Buddy's deeds, so to speak, as a fellow Korean War veteran, and he becomes for Sam the agent of a revelation at his roadside stand "between nowhere and nothing much." It is here that Sam gains access to a set of additional truths about his father's military and postwar past. When Sam asks Wesley whether

Buddy killed anyone in Korea, Wesley replies that "they don't give out those ribbons for hiding in your foxhole." Weary of the local public discourse, Sam repeats the catechism, "Yeah, my father was a hero and my mother was a saint." In a seeming afterthought on his mother's ostensible "sainthood," and further fueling Sam's inquiry, Wesley allows as how there was "that other one." Sam's head drops, registering some kind of shock; he is puzzled, queasy, but determined, as he heads to San Antonio to retrieve his father's personal effects left by Sam in his ex-wife's garage. We will learn, along with Sam, that his father had begun an extramarital affair with the young Mercedes sometime after her husband had been murdered by Charlie Wade. That affair led to the birth of her daughter, Pilar.

Central to the question of borders is the question of propriety and property, says Rey Chow in her introduction to *Writing Diaspora*. Sam Deeds has finally broached the borders of propriety and property when he discovers his father's cross-ethnic infidelity. And what that revelation unmasks in its wake is Sam and Pilar's incestuous romance, which has just recommenced all these years later. The adolescent sweethearts were half-brother and -sister. With Sam's and Pilar's adolescent ignorance now revealed, the film openly begs a rereading of the past that might sanction their erotic desires as something other than a "revenge of the past on the future," as Adam Phillips has speculated in another context (*On Flirtation*, 155).

Sayles's cinematic conclusion takes measures to reassure his audience as they confront their own taboos. Echoing the opening set on the disused rifle range, in the final scene Pilar and Sam are sitting on the hood of his car, in front of the defunct drive-in movie theater they used to frequent as teenagers. Here, where Buddy tracked them down years ago to separate them, Sam will now show Pilar the photo and love letters he found among his father's things—a photo of her mother, Mercedes, and his father, Buddy, a young smiling couple in bathing suits. He asks Pilar about the exact date she thinks her father died; he realizes he must tell her all the truth he has relentlessly sought out. She fears he will no longer want to be with her; he assures her that he is as in love with her as ever. Pilar tells Sam she is no longer able to have children, "if that's what the rule is about," she wonders. Holding Sam's hand, the history teacher offers a resolution for how their connection must override the conflict forged in their family secrets and in the founding of the state of Texas. Pilar says to Sam, in a tone of proposition, promise, and with the intonation of a question: "Forget the Alamo"—the last words of the film's dialogue, before the lyrics "I want to be a cowboy's sweetheart" cover the closing credits. The question then becomes what kind of civilization Pilar and Sam are in a position to inaugurate as both lovers and half-siblings? How else would cultural (re)production begin but with the mixing of intimates? Sayles would have us know the history of transgression, and acknowledge it, so that some new narratives can take shape at the borders of our skin and our states. Incest demonstrates a lack of respect for borders. While there may be only a line in the sand, and not yet a concrete border between Frontera and its Mexican conjoined town of Ciudad Leon, people live as if there were. Anglos, Chicanos, Mexicans, and blacks have separate interests and gathering places.

If Pilar's is the voice of a new history, Otis's amateur museum is a new kind of archive. When his visiting grandson expresses surprise about the family's mixing of black and indigenous ancestors, Otis, speaking as clearly for Sayles as did Pilar, tells his grandson—that future generation: "Blood only means what you let it."

Sayles is a director consistently concerned with lines of demarcation and power as they operate in specific communities. This is true whether we recall one of his first films, *The Brother from Another Planet* (1984), or a more recent one that takes up similar issues and is set in the *Sunshine State* (2002) of Florida. He is challenged and invigorated by the complexity of American culture and its stew—"this damn *menudo*." The challenge he issues to his spectators is to take part in the battle for a complex presentation of the issues that *Lone Star* dramatizes. Sayles demands that we contemplate stories of community, family, and kin, where blood becomes what we are able to let it be or not be. Sam and Pilar will reconstruct a past in which their father, Buddy Deeds, is neither beatified nor demonized. While an embodiment of the law of his land, his love led him to cross borders—marking an end to an era of blood shed, ushering in an era of bloods mixed. It is up to Sam and Pilar, in their moment of mid-1990s multicultures, to arrive at some new names for the relations they have and will forge. Their exchange in the final scene opens a debate on love and the law.

Civilization and its founding "deeds" encompass founding taboos. For commandments to be delivered—that which must not be done—those deeds must have been done, and to excess; outlawing them founds what becomes the law as it is lived and written. Indeed, blood and its movements across bodies, families, and nations remains one of our fundamental civilizing myths and juridical forms. Sam is out to discover what blood his father might have shed. What he finds is not homicide but the mixing involved in an adulterous and interethnic romance. Sam is a postoedipal hero in the terms of Western foundational narratives in Sayles's epic of the refounding of a civilization in the borderlands, where, as Freud noted, concluding *Totem and Taboo*, "in the beginning was the deed."

3 THE NEWLY BORN CENTURY

Can we summon up a sense of exuberance we once may
have felt about the century we are now living in? Can we
recall our immense curiosity about beginning to weave
ourselves into a new time? Can we recollect an eagerness to let it permeate our daily lives with a sense of
the unprecedented?

The demographers collect narratives indicating that
the centenarians among us are on the increase. New
centuries carry energies that proliferate stories of birth.
What would it mean to consider our new century as a life
cycle narrative—developmental. We are early in it still.
Our decade not yet baptized—nicknames, pet names
proposed, but none has stuck. Some cultures await the
emergence of the personality to name the child fittingly.
What else do we intend when we characterize decades,
generations, centuries?

Surely we still remember the stir of energies aroused
by the coming of that shorthand marker known as Y2K.
As the year 2000 approached there was a growing sense
of anxiety that surfaced through the anticipation. And for
all the dread, even, we made it through the Y2K millen-

nial moment. The lights did not dim on our electronically based fires of civilization. But debates continued until the coming of 2001, which some claimed was the rightful beginning, or birth, of the millennium. The trauma of this century's birth would come before 2001 was out.

By early 2003, as I first read Jeffrey Eugenides's exuberant second novel, *Middlesex*, I kept thinking, wishing, hoping that we'd just passed through our terrible twos. And *terrible* was the operative word, so far. I kept wondering what we might learn if we were to address the century as *in-fans*. Un-formed, mal-formed, in-formation? Information. Surely the world has always required our care and "feeding"—and the language of the "world" had fully given way to the perspective on the "global."

My sense of the century since Tuesday morning, September 11, 2001, has been that it was the spasm of a monstrous birth whose afterbirth spun as far as Bali and whose spores we are yet learning to breathe. At the end of the twentieth century, we were becoming fluent in the metaphors of immunology. The twenty-first is insisting on our viral loads and genetic connections. The freedom I take heart in and claim for my own purposes here is the play between narrative and knowledge, stories of science and stories of family, stories of tribes and stories of strangers. The philosopher Anthony Appiah, in his recent book *Cosmopolitanism: Ethics in a World of Strangers*, takes conversation as a model of how to engage the "global tribe we have become" (xiii) and points out that "evaluating stories together is one of the central human ways" (29) we can approach each other through our mutual, contemporaneous differences. Where I speak in praise of mixing, Appiah goes further in his "praise of contamination" (111) as a way to recognize our potential porousness to strangers.

So here, in the babbling pre- and postoedipally stammering years of this century, what might we decipher and discern about optimal conditions to thrive? How can we speak for our time with all we've learned about the problems of speaking "for" an other? But this newly born time insists that we engage. And it appears to be an *enfant terrible*, this cursed twenty-first-century turn. We must work harder than ever to love it, or it will just keep thrashing about, struggling over boundaries to be breached, set, and respected. The terrible infant must be so because it does not articulate—semiosis doesn't simply sing; it stammers, screams, stutters, babbles, coos, calls, and cries.

The opening words of Eugenides's novel *Middlesex*—"I was born twice: first, as a baby girl . . . and then again, as a teenage boy" (3)—are spoken by Calliope/Cal Stephanides, a prenatally omniscient narrator. In this new time, seemingly stable documents, like birth certificates, no longer reliably certify attributes such as name and, in this case, sex or gender identity. What are the identifying markers of this child? If Eugenides is to be our narrative guide to this newborn babe, we need new answers to old questions. The first question at birth isn't even about wholeness—it's about halfness. It's said it's the fingers and toes that count. Yet isn't the classic first question: Is it a boy or a girl? And from there on out it gets sticky with sameness and difference: color in those eyes, that hair, that skin; shape of that nose, chin, forehead. The bio- and the social weave their postnatal tapestries.

To backtrack a bit before moving forward: in 1996 Jeffrey Eugenides published a short story called "Baster," a tale of revenge told by a nameless spurned man. It was a decidedly low-tech parable of the mating scene in New York City in mid-1990s America. The narrator switched his own sperm sample with that of the man who had been chosen to father the child of a woman hoping to become a single-mother-by-choice. Chance and fate intervene otherwise in *Middlesex*. Genealogies of genetics are scripted through an epic Greek immigrant saga in which the burning of Smyrna in 1922 and of Detroit in 1967 draw the reader from the beginning of the twentieth century to its end and into the twenty-first. This is a hybrid time told by a hermaphrodite, Calliope Stephanides.

Eugenides certainly writes in praise of mixing it up—the crazy gene pool we swim in. And getting under each other's skin is certainly our only hope if we wish to save our own and each other's lives. His concerns mirror mine, though his take form in fiction. I want to use his fictions to ask questions about the facts that come to make up family. In Eugenides's move from "Baster" to *Middlesex* he ranges from the home-pregnancy kit and whole-earth mixing of gene pools to the diagnostic genetically tested answers to the riddles of intersex and gender.

Even his titles tell this story: "Baster"—kitchen utensil conventionally required for home insemination, as we imagine done in the private domestic spaces of determined nonheterosexually reproductive women in the closing decades of the twentieth century; *Middlesex*, with its Anglo-Saxon resonances, asks us to take up the intersex questions that come to usher into the world his "prefetal" narrator, who will tell us that s/he was "born twice." For Eugenides the epic *is* also the genetic. A three-generation Greek family saga overdetermines his Homeric tone: "Sing now, O Muse, of the recessive mutation on my fifth chromosome!" (4). This is a family founded on the "marriage" of siblings fleeing for their lives from the massacre and burning of Smyrna in 1922—one of the earliest of the genocides for which the twentieth century necessitated the coining of that word.

Just as we may not get to cousins without passing through siblings and their incestuous intimacy, consider the obvious fact that genocide—the mass effort to annihilate a people connected genetically (by blood in the commonplace metaphor)—necessitates, requires, insists that its survivors start over in the face of loss. Remaking kin, remapping genealogy will, of course, emerge between those who are able to entrust their lives to each other. As anthropologist Annette Weiner puts it, "The established theories of descent rules and marriage exchange ignore the social prominence of sibling intimacy as a basic kinship principle" (*Inalienable Possessions*, 151).

In the mountain village of Bithynios, where *Middlesex* begins, "everyone was somehow related" (39). Third cousins frequently marry, thereby rendering siblings also already cousins. The desire that powers Lefty and Desdemona's sibling kiss will stir the gene pool that two generations later will bring us their grandchild narrator, their granddaughter/grandson, Calliope/Cal—who is also their great-niece/great-nephew. Eugenides deftly leads his readers into this moment of meeting, mating, and determining life, history, and family events as the Turks drive the

Greeks out of Asia Minor and, in this particular case, over the Atlantic and ultimately to Detroit.

Our narrator's real concerns while telling this gripping, comical, poignant family epic are the questions of the relation between chance and fate that always track families across time and space. Let's take "family secrets" to be redundant, perhaps even definitive of the cauldron that is family. The two who live the secret of Lefty and Desdemona's sibling coupling and marriage are soon joined by the required third for the circulation of secrets. Their secret of incest is laid bare much later as one of genetics—a veritable epistemological field of secrets revealed. This tie between Lefty and Desdemona, each the other's object of desire, echoes through the double helix of generations. Two generations later Eugenides asks us to listen to Calliope, the muse of epic poetry, as she sings the tale of her grandparents' childhood and courtship. Some family secrets, like this one, go underground into the familial unconscious, but "what humans forget, cells remember" (99).

Greeting the newly married immigrant couple in Detroit is cousin Sourmelina, who had immigrated earlier in the century. She is already married and longing for a child; the two wives will share their pregnancies and their deliveries: one a boy, Milton; the other a girl, Tessie. And a couple of decades later, having "looked up all the statutes" (195), these second cousins will marry. Carrying one gene mutation each, they will have two children: their son, who will always and only be known by the name Chapter Eleven; and their daughter, Calliope. As if in inverse proportion to the siblings whose intimacy began this line of kin, this sibling pair exemplify the potential distance the biosocial can also create.

Eugenides's hybrid creation of Cal/liope is shot through with the tropes of classical literature and writing. After all, chimeric beings populate mythologies and bestiaries of old. Chimeras: peculiar combinations of animal-like creatures who are meant to allegorize human aspects, characteristics, behaviors. The human is fully implicated in the bestial, the bestial in the human. Cal/liope's twinned self—she will ultimately become her own opposite-gendered sibling—lies coiled within her being and is only revealed slowly to her and her family. At adolescence, an awkward, chimeric self-state of neither and both child and adult, Calliope recognizes, with a healthy dose of self-love, that she is not quite like her girlfriends: she is a changeling; she has that "saluki look" (304), and since it's the 1970s and androgyny and skinniness are trendy, she is not subject to the primitive taunts by peers who "call them tomboys or worse: ape-women, gorillas" (304). By age thirteen Cal looks back on herself as an "awkward praying mantis of a daughter" (305). In the next few years s/he will fall in love with the girl known only as the Object, and the "fugue state" of their "early sex" brings Cal/liope great pleasure as the "body, like a cathedral, broke out into ringing" (386–87). Her passionate love for the Object will exceed and supersede her splitting of gender and sex at these critical moments of adolescence: "Through all this I made no lasting conclusions about myself. I know it's hard to believe, but that's the way it works. The mind self-edits. The mind airbrushes. It's a different thing to be inside a body than outside" (387). It is only when an accident lands Cal/liope in the emergency room, and medical discourse begins to problematize her body, that the "crocus" of her sex that s/he thrilled to is designated aberrant. And the entire family energy is marshaled for a trip to faraway New York City to consult with the proper authorities. There, her parents attempt to maintain the equilibrium of their love for their child, even as dire medical reports are shared with them to which Cal/liope is not privy. Cal/liope takes off for that great people's resource center, the New York Public Library, only to read in the biggest dictionary s/he has ever seen, the definition of that word overheard, *hermaphrodite*, to find that the entry ends with "See synonyms at monster." "But the synonym pursued her. All the way out the door and down the steps between the stone lions, Webster's Dictionary kept calling after her, *Monster; Monster!*" (430, 432). Callie's father, Milton, tries to explain both to himself and to his daughter that "the bad news is you have to have a little operation" (432), and fleeing the medical expertise of the 1970s that recommends surgery to "correct" her intersexed nature, she becomes a runaway teenager, ending up in countercultural San Francisco. There her body, and her estrangement from it, will lead her into marginal work in the Tenderloin, where she dances underwater, being seen without seeing those who ogle her—coming to learn femininity even as she discovers her own masculine self— a transgendered being, long before that language emerged in public culture. Mary Shelley told the tale that moves us still of what happens when we don't love our monstrous offspring. Though Callie's references for the monstrous run not to Frankenstein's creature, but Bigfoot or Loch Ness, s/he is a character who teaches, as did Shelley's, that we must let our monsters out—they demand and deserve recognition—they are us: our same, self, others. Eugenides's tale is one of ambiguous gender and its ramifications across time and of how we, his readers, will tolerate such ambiguity. Even as I write this in late 2006, certain states have begun to pass legislation allowing sex designations to be changed on those once stable, even rigid, documents like birth certificates. And on December 2, 2006, we read and learn, starting on page 1 of the *New York Times*, about gender identity, gender variance, and transgender—our latest renderings of what was once thematized by those earlier Greeks in Tiresias's story of having been both man and woman.

Recognition is a key moment in classical drama—*anagnorisis*, it's called. And it often accompanies tragedy, most famously when Oedipus's parricide and incest are followed by his self-blinding. But as I said earlier, Eugenides is writing an exuberant fiction, powered by a hope for the capacity of the new century to tolerate ambiguity, contradiction, the copresence of opposites. And Callie will also come to a moment of recognition—several such moments, in fact. Just as a medical emergency led to new and difficult self-knowledge for Cal/liope, the next crisis is the announcement of her father's sudden death in Detroit, necessitating a return to the family of origin; they have not laid eyes on Cal/liope in a few years. Grandmother Desdemona is on her deathbed, waiting to confess to this young man she barely knows as her granddaughter the truth of her incestuous marriage nearly a century earlier; Chapter Eleven remains nonplussed and in character, not quite knowing how to absorb the fact of a brother—no longer a younger sister. Speaking for Tessie, Callie's mother, Cal clarifies: "It was not acceptable that I was now living as a male person. Tessie didn't think it should be up to me. She had given birth to me and nursed me and brought me up. She had known me before I knew myself and now she had no say in the matter" (519). Of course, in running away Callie had made a decision that would change her life as dramatically, if not as painfully, as the surgery the expert was recommending. And s/he knew that: "in the end it wasn't up to me. The big things never are. Birth, I mean, and death. And love. And what love bequeaths to us before we're born" (388).

What love has bequeathed Cal/liope is the capacity to tell this story of mixing—mixing genes, mixing genders. The novel closes with Desdemona's blessing to the prodigal grandchild that when she is gone, "you can tell everything" (528). The change of genders doesn't appear to perturb her wise, old self. The ancient and the postmodern are intimates here, as Cal, now a man, stays home from his father's funeral to keep Desdemona company, to avoid having to explain herself/himself to old friends and family acquaintances all in view of Tessie's still somewhat shamed maternal recognition, and finally, to perform the traditional function of a man in the Greek Orthodox family: to guard the door "so that Milton's spirit wouldn't reenter the house" (529).

If Eugenides's novel and its main character are templates for "what was next," then we have to conclude that the generation of Cal's parents, the ones who occupy the central focus of the twentieth century, are the ones who have the most difficulty contemplating Cal's very "nature," while he and his grandmother have the insights of the past to guide them as the extravagant possibilities of the future come hastening toward us. Cal and his grandmother find and relish the strangeness in that which may be most familiar. And in hearing each other out, Desdemona on her deathbed, Cal at the beginning of his second life, they change and cancel out dialectically some of the rigidity of categories of identity undergirding much twentieth-century thought about selves, nations, and their differences.

4 SISTERS OF THE BONE

Anthropologists bring us stories about the myriad forms taken by the urge to create systems of connectedness, commonly called kinship systems, among humans. The variability with which people have arrived at the ways of reckoning their near and dear ones is astonishing. Yet in 1997 in French, and in 2001 in English, a Chinese anthropologist who studied in France but did fieldwork in China presented what he considered to be a unique—in fact, what he called a "unary"—form of kinship. This term is opposed to the "binary" forms that emerge from what may vary widely but is understood by the term *marriage*—that proximity of care that derives from what is usually a two-some acknowledged through a set of public ritual acts. The people studied by Cai Hua, an anthropologist, are the Na of China. For the Na, what is unique is indicated by Hua's title, *A Society Without Fathers or Husbands*.

Among the Na we learn of a culture where the "furtive visit" is the usual mode of sexual intimacy: men visit women's homes under cover of night. Women don't visit men. Furthermore, these visits must not be observed by any others living under the same roof. Coexisting with the furtive visit is what Cai Hua calls the "taboo on sexual evocation." These visits are known but remain unspoken. Households are constituted by siblings, their mothers, and her brothers (i.e., uncles to the younger generations). The sibling relation through a mother line, or *matrilinee*, is the way kinship is calculated, understood, and organized. And it is bone, not blood, that figures in the naming of kin. Children of the same mother (line) are said to be of the same bone. Now the furtive visit is not a singular occurrence; young people may make and receive numerous visits in a given night. Lovers may last a night or even a year, but they remain private, or furtive, in their activities; yet they are understood to be connected. Lover relations, we learn, are not coercive. Jealousy is considered shameful. Anyone may choose to drop or take another lover at any time. If one is rejected, one is expected to accept this without struggle or conflict. Laying bare the anthropologist's own confusion over this situation, Hua explains that "a man's sexual life takes place away from his daily life. 'Acia' are lovers who remain strangers to each other. The acia relationship is . . . a purely emotional, amorous, and sexual relationship. It is a private matter" (*Society Without Fathers*, 232). Clearly legible is Cai Hua's effort to resituate the lovers comfortably within structures more familiar from his training at the very home of mid-twentieth-century structural anthropology—the Sorbonne. Claude Lévi-Strauss is notably among those asked to blurb Cai Hua's study, and he acknowledges the importance of this work on a society that would "deny or belittle the roles of father and husband" (dust jacket of English translation). One may read in this statement the insult to the models more familiar to Western readers.

We should perhaps not be surprised that others term the Na a matriarchal society; in fact, during 2006 *The Women's Kingdom*, a twenty-minute award-winning documentary about the Na, was screened at film festivals. The director, Xiaoli Zhou, who was born in Shanghai but was a student at Berkeley, wanted "to see for herself how much freedom a woman might enjoy in China." For those who have visited and documented the Na (called by others the Mosuo), there is a clear urge to let the world know about and try to understand what may indeed be a culture that is vanishing.

Why do I resort to this story for this taxonomy of what we need to learn to read in the realms of intimacy? Because we live in the midst of a sea change about kinship, and we need insights from those who have perhaps made very different but equally long-standing choices, ones that indicate how very differently we can learn to think about how we designate those for whom we are bound to care. Here is a world without words for husband or father—or cousin. It is the anthropologist himself who, to accomplish his fieldwork, asks his native informants whether individuals know who, among the men in the surrounding households and neighboring villages, might be anyone's "genitor." This is the term to indicate a biological connection that appears to mean nearly everything to cultures such as ours and nearly nothing to the culture he is studying. The replies he gets indicate that some people know who their genitor is, but that it doesn't matter; nothing of their status is affected by such knowledge or the lack of it. More surprising even is the baffled reply to a further question of Cai Hua's; when he asks whether someone knows the genitor of their genitor, he is met with outright derision. "We never ask or think to ask such a question! What would this mean?" (228). So-called consanguineal (sharing blood, not bone) relations through the mother line can be followed through female ancestors without any difficulty at all. This is what counts for the Na. Sisters and brothers are the primary unit of kinship, and this is true across generations. This does not, however, denote the absence of a gendered relation to labor, nor is it the destabilization of patriarchy—note that women may not visit men without risking their reputations and that brothers and uncles tend to organize household duties and responsibilities that involve the outside of the home. It is, however, tantamount to an utter lack of concern with paternity, a foreign concept. How, then, do the Na understand sex, in which they indulge freely from their late adolescent through adult lives? It is presumed by proverbs and other kinds of statements about sexual activity that in having sex with women, men perform an act of "charity," compared to the rain necessary to make the grass grow. To that extent, their contribution to reproduction is acknowledged.

The profusion of contemporary anthropological work on kinship—much of it from a feminist perspective—reiterates the dialectical mix of affinity and alliance, usually also calculated through a bodily substance with the weight that brings to sentiment. Janet Carsten's succinct formulation at the level called the consanguineal insists in many cultures on blood and sometimes on seminal fluid. Among the Na, *substance* and *sentiment* are held to be in the bone. Affinity and alliance make up the material by which the discipline of anthropology has established its terrain. The expected shock effect of the Na system of kinship is repeatedly apparent in Cai Hua's discourse. He acknowledges that the Na model is based on a form of consanguinity, an emotional economy requiring that there be descendants of a household, and one where the brother–sister bond is, as it were, sacred (465).

The Na teach us to think further about the sibling relation for its model of continuity, support, and care. As our kinship paradigm moves increasingly toward a genetic understanding of differences and similarities, we need alternative frames of reference as we rewrite the terrain over who lives together, how they love, and what is meant by the sharing of substance, whether blood, bone, or genes. Our model redesigns the scientific through the social helix of fate and choice. I imagine that as our knowledge grows in cutting ever finer distinctions through genetics, what will challenge us are the stories passed through science that demand new forms of literacy. We may learn to think relatedness backward and forward in time and history that will undoubtedly come to include ancestors and concestors. The sophistication of genetic testing down to mitochondrial levels of connection to our primordial Eves and Adams is increasing exponentially each year. We are learning rather rapidly to recognize the degrees to which we are, all of us, mixed—as we simultaneously rethink the basis for and of our common differences.

In his review of Cai Hua's book, Clifford Geertz revisits the efforts of various programs in postrevolutionary China and their failures at getting the Na to take up some form of state-sanctioned marriage or even cohabitation. Geertz concludes by saying, "In China, as elsewhere, it is not licentiousness that powers most fear. Nor even immorality.

It is difference" ("The Visit," 6). The tourism industry has come to the region inhabited by the Na, and young Na women have entered into prostitution, reports Tami Blumenfield; this has evoked conflict with their female elders who are unhappy with the public nature of this "sex work" and with the notion of wages. With a population estimated to be thirty thousand and dwindling, and with the coming of roads followed by tourism, and the lure of city life elsewhere, it is not farfetched to see the Na as a vanishing human species.

As animals we seem to make our marks based on proximity in space and time by virtue of the ground on which we have grown. There is a "circle of empathy" that may begin very locally, but desire leads us to draw the circle more widely. Intimacy draws us out beyond the circle, but not too far out, while simultaneously demanding we stay close, but not too close in. Among the Na the need for descendants brings about a form of adoption of others into the matrilineal household so as to assure the elders of younger members who will care for them into the future. When possible, these adoptees are solicited from households that may have some kind of geographical or historical connection so that it is still possible to conceive of those of the "same bone" continuing to create and care for those who become members of their households. As seen in the furtive visit, these homes welcome outsiders. They come in the dark and are not spoken of, even as they expand the circle of empathy, or the genealogical realm, without any of the desired forms of regulation expected by the dominant Han Chinese cultural practices.

Cai Hua, through his acknowledged hegemonic Eurasian prejudices, examines the kinship structure and its elaborate effects. Where he fails to pose questions is in the affective realm. He seems only to want to name the structures so foreign to his ears and, he assumes, to ours. For example, we get a treatise on a "society without fathers or husbands," but this is certainly not a society without children. Yet his focus, and therefore ours, is on an absence—the husbands and fathers. And the place children have in this society also remains blank after some five hundred pages, except as they function to ensure continuity and elder care. If there is no sentimentality between lovers, what does intimacy look like? For we also do not get a full sense of the nature of the attachment between siblings and their elders. But particular anecdotes tell suggestive tales.

Cai Hua's ethnography is found wanting by several commentators and reviewers; as someone outside the discipline, what I glean in these critiques is praise for his masterful structuralist study, which nevertheless lacks the postmodern ethnographic narrative that would give readers a sense of what it is 'to be a Na." But as a Han Chinese field-worker, Cai Hua would, I imagine, move gingerly in an effort to keep his work from being used to prescribe, as various governmental efforts have, and simultaneously not to have his work serve the ends of sex tourism either, avoiding any tone of prurience in his questioning about the furtive visit. If, as Geertz claims, written knowledge in the West about the Na goes as far back as Marco Polo's *Travels*, then Cai Hua surely is attempting not to sensationalize or to have his work underwrite the travel fantasies of domestic and foreign tourism. Cai Hua is not the only anthropologist to be working among the Na in recent decades.

Others, such as Tami Blumenfield and Eileen Walsh, aim to fill in the paucity of gendered perspectives, as does Zhou's film *The Women's Kingdom*.

We learn that when young women begin to go out (the movies are still a relatively new place for social life to occur), if they engage in conversation with young men, then they are also presumed to welcome displays of physical affection—touching that may lead to a later furtive visit, and perhaps to becoming "acia," or lovers. The more recent women visitors to the field have called these relations "walking marriages," naming their fact as opposed to their form, as does the "furtive visit." While the term *marriage* suggests duration, even permanence, the furtive visit opens the terrain of multiple sexual liaisons, while also clearly judging some aspect of this practice as unassimilable. Like Cai Hua, women anthropologists recognize the rarity of a culture organized through women's desires; however, unlike him, they sound notes not only of enthusiasm but also of alarm about the endangered status of the Na, or Mosuo, people.

There is heteronormativity at the core of Hua's ethnography, and there are forms of patriarchy operating among the Na. Yet, given those limitations, what is striking are the forms of a society where marriage is not institutionalized and where sexuality is integrated with what seems to be an attitude that assumes pleasure, all regulated by the desires of women young and older. Forms of sexuality that suggest a lack of coercion and that are practiced with a regularity established by individual desires are allowed to flourish in a household organized through mother lines. Cai Hua claims to find only one instance of a promise of fidelity between lovers; such vows are considered shameful because sexuality is not a currency to be negotiated among partners; it "implies no mutual constraints" (215).

Closer to home, in June 2005 you may have read about two women who fled a fundamentalist Mormon town where plural marriage is the norm and often there is coercion of young girls into marriage with sometimes much older men. These women challenged the patriarchal norms in the governing of this community, and they made the "news" because the upshot of their struggle was that they had won the right to become members of the governing board of this isolated town. Their example has fueled the energies of other women in their community to begin to speak out about their own oppression. Or you may be following the popular cultural plot of the HBO series *Big Love*, in which both red- and blue-state Americans are being put to some narrative and cultural tests if they are to identify with the women and men in this tale of plural marriage. As the second wife of three has recently commented about her diminished circle of empathy, "I gave up 100 to get 10," as a way to quantify her loss in her move from a fundamentalist community, with its isolation, into a suburban yet secret life that promises more autonomy even as it borrows from and revises the traditions in the more "alien" community that both this wife and her husband have fled. But in these communities of sister-wives and multiple mothers, more than two sets of grandparents and siblings whose ages may approximate those of the newest wives, and exponentially expanded kinship networks, we also see that such intimacy, proximity, and isolation produce rich and thick conflicts *and* connections.

In our current political debates over same-sex marriage, we can see another struggle over numbers and differences;

there is both too little difference in the locution of "same" sex—there must be *one* man and *one* woman—and too much in this demand because opening up the definition of marriage is seen to be, and most likely correctly, opening the door to multiple partnerships. As I've noted earlier, convention calls for a move toward difference—away from incest—and simultaneously a move that hovers near the same—away from strangers, or simply strangeness, that which we know little of. With the Na, as with the Mormons in earlier times, the state has tried unsuccessfully to intervene in notions of kinship, to enforce dominant heteronormative habitation and marriage patterns. Fundamentalism of any stripe tends to reflect all the worst aspects of patriarchal control of women as property. Canada, our closest cultural neighbor, has acknowledged the rights of gay men and lesbians to marry with relatively little public struggle. These forms of kinship remade through love and laws are coeval, contemporaneous in history, and their ability to coexist needs to be understood as a sign that we are able to tolerate so much more difference than we are led to believe by those who wish to police the borders of intimacies. If societies are constructed on and out of contradictions, then the Na taboo on sexual subjects, coexisting with the institution of furtive visits, asks us to take account of some of our own cohabiting oppositions in the realm of kinship, our determinations about those we keep close and those we take to be strangers. Wouldn't our efforts to cooperate rather than collide be furthered by the notion that we are like our nearest others and they are like us? They and we are able to adapt to many forms in our yearning to keep those we care for near.

In considering the Na system of kinship, I want to make of this one example a case for readers to consider how variety comments on and inflects our own prejudgments about how intimate life may be organized. At a time when struggles over gay and lesbian forms of family and childbearing and rearing are flashpoints for reactionary discourse, I propose a wider knowledge of how a sibling model of relation begs us to move toward more mixed models of intimacy and more tolerant ambiguously gendered structures. It is time to come out from our caves of thinking, fearing, feeling, and protecting to see what other intimate life forms are possible—to shed our fear of numbers, small or large.

Finally, I would ask readers to follow my move from a vanishing, endangered human culture across the thinnest human–animal line. What might we learn from the explosion of the concept of our kinship, our cousinship, across primate lines? Not just from the chimps but from the still lesser known yet more closely related cousins, also vanishing rapidly in the wild: the bonobos, where females also regulate habitation and sexuality.

PART II

That touch of the baboon all babies have.

—Jeffrey Eugenides, *Middlesex*

Linnaeus, the founder of modern scientific taxonomy, had a
weakness for apes. . . . [H]aving returned to Sweden and become
the royal chief physician, he gathered together in Uppsala a small
zoo that included various species of apes and monkeys, among
which it is said he was particularly fond of a Barbary ape named
Diana. . . . In a later writing bearing the title *Menniskans Cousiner*,
"Man's Cousins," he explains how difficult it is to identify the
specific difference between the anthropoid apes and man from the
point of view of natural science.

—Giorgio Agamben, *The Open*

5 APES 'R US

> ... the monkeys stand for honesty, giraffes are insincere,
> and the elephants are kindly, but they're dumb
> orangutans are skeptical of changes in their cages
> and the zookeeper is very fond of rum ...
>
> —Paul Simon, "At the Zoo"

It is a commonplace that the characteristics we apply to animals say as much, if not more, about who we are than about the kindred creatures themselves. Human attributes offer a template onto which we can project our resemblances and our differences in ways not necessarily distinct one from the other—what we see is mirage, mirror, and more. Humans make boundaries between themselves and the other animals with whom they share the planet. In that very marking of territory we draw out their differences from us, even as we attribute to them what we are, what we wish to be, what we dare not be. We make of them portraits of the selves we desire and deny. We anthropomorphize with delight and with horror, with abandon and with reserve.

Bestiaries, among the oldest of literary genres, are ever ready to be made new again. They are fables of the human–animal borderlands: where the lion becomes king, in a creaturely cosmological network. Connections across species boundaries are currently transforming the stories we tell to help us speak these interrelations. One of the central dynamics in the human–animal mirror stage, so to speak, is between what we consider *natural* and what we assume to be *cultural*. There is a literature both scientific and popular, and growing exponentially, on our commonalities with nonhuman primates. Cross-species medical technologies, transgenic identities, and other chimeric beings force us to rework, rethink, rewrite our epistemological categories. The last time these relations underwent such thorough upheaval by new knowledge was during the nineteenth-century expansionist era of colonialism and imperialism. While much of this knowledge is no longer new, we struggle mightily with its implications in this new century, with some of our conceptual guides the thinkers of the last fin-de-siècle. These "hot zones" of contact, connection, and conflict between humans and our closest animal kin reward contemplation and speculation for the ethical and intimate issues they raise. This new millennium has made it entirely obvious that scientific illiteracy threatens our ability to keep pace with ever-changing and shifting debates, from those surrounding stem cell research to the questions around how reproductive technologies ought to be put to use.

Current social conditions have formed kinship networks that demonstrate the plasticity of the realm of intimacy and imagination, even as we simultaneously see how vigilantly some guard this emotional terrain. Frans de Waal's most recent text of public scholarship comes out of a series of lectures he gave at Princeton in 2004; they are collected in *Primates and Philosophers*, along with several critiques from the community of scholars dealing with questions of ethics and empathy across species differences. In those debates de Waal calls for "intellectual breathing room" that is gained by positing or assuming that "instead of empathy being an endpoint, it may have been the starting point" (23). Like other primatologists who turn to the mother–infant bond to hypothesize this experience as our primal encounter with conflict and reconciliation in the act of weaning, de Waal speculates on what some call "proto conversations" between mother and child (23). In the experience of weaning, those who have been most intimate must establish a degree of difference, a move from nearest proximity to appropriate distance. There we gain emotional, empathic knowledge about where former intimates become necessarily estranged. What if those same capacities, newly inflected, allow us in turn to make intimates of strangers? What enables us to see ourselves in the mirror of another's face? Isn't the spark of recognition ignited by the fleeting apprehension of sameness in difference? Thinking about kinship across different species in his recent book, *The Ancestor's Tale*, Richard Dawkins turns to another ancient genre of writing—poetry, specifically *The Canterbury Tales*—in order to spin out his reading of evolution backward. Why not take seriously, very seriously, the fact that although we are human, with all that allows and constrains, at the heart of what that means is that we are animals. We are the "naked" apes of the world, and from the beginning we have drawn on our animal kin to allegorize ourselves—hence the bestiary, which along with the Bible could be found in the earliest libraries. In it we read of all sorts of animals, both imaginary and real, and we learn of their habits and customs as ways of engaging our own place in the world. After all, we are the ones who tell stories of them.

Here are two such stories from Richard Barber's *Bestiary*, which is translated from the Latin and organized from lion to dolphin, passing through phoenix and hedgehog, among many others. Consider the beaver, for example, from description to allegory:

> There is an animal called the beaver, which is quite tame, whose testicles are excellent as medicine. The naturalists say of it that when it realizes that hunters are pursuing it, it bites off its testicles and throws them down in front of the hunters, and thus takes flight and escapes. If it so happens that another hunter follows it, it stands up on its hind legs and shows its sexual organs. When the second hunter sees that it has no testicles, he goes away. In like fashion everyone who reforms his life and wants to live chastely in accordance with God's commandments should cut off all vices and shameless deeds and throw them in the devil's face. Then the devil will see that that man has nothing belonging to him and will leave him, ashamed. That man will live in God, and will not be taken by the devil, who says: "I will overtake, I will divide the spoil" [Exodus 15:9]. The beaver (castor) is so called because it castrates itself. (43–44)

More to my purposes here in terms of naming the beasts is the entry on the ape, also taken from this text, layered through the Greek *Physiologus* and "preserved in a kind of Christian aspic," to quote its contemporary translator and editor, Richard Barber:

Apes are so called because they ape the behaviour of rational human beings. They are very conscious of the elements, and are cheerful when the moon is new, and sad when it wanes. Their nature is such that if a mother bears twins, she will love one and hate the other. If she happens to be pursued by hunters, she will clasp the one she loves in front of her and carry the one she hates on her back. But when she is weary of running upright, she willingly drops the one she loves and unwillingly carries the ones she hates on her back. Apes have no tails. The devil has the same form, with a head but no tail. If the whole of the ape is hateful, his backside is even more horrible and disgusting. The devil was at first one of the angels in heaven, but he was a hypocrite and deceitful and lost his tail. . . . "Simia," the Latin word for apes, comes from the Greek and means "with nostrils pressed together." Their nostrils are indeed pressed together, and their faces are horrible, with folds, like a disgusting pair of bellows; she-goats have the same nostrils. Monkeys have tails, but that is the only difference between them and apes. There are also baboons, which are very common in Ethiopia. They can make great leaps, and their bite is severe. They can never be properly tamed, and always remain rather wild. Sphinxes are also a kind of ape, with shaggy upper arms; they can be taught to forget their wild nature. (48–49)

Notice those omnipresent hunters; they are the ones bringing back news of severe bites, horrible faces, and disgusting backsides as they engage in pursuit of others who must be excluded from our own God-given purposes and designs. Notice also the projection of child-rearing practices and negotiations of care and conflict under the pressure of depredation by humans. And especially notice the directionality of the mirroring of ape and human, the uprightness and protection of younger vulnerable ones of their own kind. Certainly these words testify to what Nato Thompson calls our "monstrous empathy," which channels the ways that "animal/human relations have been ensconced in the repulsive longing of taboo" (*Becoming Animal*, 8–9).

Apes 'r Everywhere

In Franz Kafka's short story "A Report to an Academy," an ape narrator addresses humans. My own position here echoes such an encounter as I retell the stories I find circulating among those in the world of primatology in order to bring them into the literary critical languages of cross-species common knowledge. I translate—historically a thankless task—across disciplinary lines with a full sense of the risk that bears. What can we learn to see when we look through the bars, across the moat, into the eyes of our others, our betters, our kin and kind? What lives there of conflict and connection if we dare to look, to learn, and to listen?

The world of primatology is a very busy and exciting locus of new knowledge. While primates and primatologists who study them have made the news repeatedly in recent years, their appearance in the realm of literary criticism may still be rare for what the young interdiscipline of animal studies may offer those of us who study stories and storytelling and storytellers. In the arena of public culture, I still find many who have not made the acquaintance of the rather recently renamed primate, known as the bonobo, formerly called the pygmy chimpanzee. Like the Na people of China, discussed in chapter 4, the bonobos are, to quote Frans de Waal, a "hot topic about a tiny community" (*Bonobo*, 172) through whom we have much to learn as subjects.

Among these "great apes," unlike among chimpanzees, orangutans, and gorillas, females are said to dominate, or to be codominant with, males. Their only habitat is the politically and ecologically endangered rain forest south of the Congo River. They have come to be known as "the gentle ape" because of their resolution of conflict through sex rather than aggression. One of their most visible spokespersons, de Waal, has called them "the forgotten ape" in a book with that subtitle. I am curious about what cultural meanings can be derived from our willingness, even eagerness, to invest with kinship (our nearly 99 percent shared DNA) those whom we have long counted on to reassure ourselves of our dominance in the world of nature and culture. Our resemblance and claim to relation is evident in the ubiquitous metaphor of them as our "cousins" in the popular literature about great apes. And the kinship category of the cousin is among the most plastic in its ability to suggest both connection *and* conflict. Here is another site (about which I will have more to say shortly) at which to look closely at the negotiations around kissing and killing that go on among kindred spirits.

As this project has matured over the past decade, nearly each week of recent years has brought news of our genetic neighbors that takes up how we are related to even so-called lowly life forms—yeast, for example—and how this genetic mapping must humble us in our efforts to understand how we remain under the sway of the biological even as we insist on the social correctives to how our lives are shaped. Biopolitics keep taking on more and newer meanings as we become increasingly sensitive to the power struggles of living confronted by other creatures and the power-knowledge struggles between the human animal and our animal others.

What are the features of their struggles for dominance and hierarchy? The primary agents in these struggles among bonobos are alpha females. That researchers could acknowledge this now and not previously suggests a multiply determined scenario of new knowledge, one that has great potential for firing our imaginations. It is no coincidence that the primatological imaginary and research agenda has been richly fed by the work of women in the field—a scientific discipline where women are among the most renowned. While Jane Goodall and Dian Fossey are familiar names, other primatologists who should be as highly regarded include Sue Savage-Rumbaugh, Barbara Smuts, Meredith Small, Karen Strier, and Amy Parish, to name just several discussed by Carole Jahme in *Beauty and the Beasts*.

In *Interrogating Incest*, Vikki Bell asks that we understand feminism as new "ways of seeing and ways of hoping" (199). We humans have been forced time and again by historical and social circumstances to reexamine our own group relations and their instability. Feminist movements—women acting en masse to bring about change in the asymmetrical structures of cultural dominance in its gendered nature—exemplify how utterly discordant and divisive such challenges are—always and everywhere. Such reexamination does not occur harmoniously.

The irony that led us through the last decades of the twentieth century and that characterized certain post-modernist discursive practices displayed a certain lack of humility, as irony will. And in some of the discourses of feminism, we can read an equally imbalanced strain of earnestness. Both critical theories led to some affective and intellectual cul-de-sacs that I would suggest were obliterated in the spasm of the new century that occurred in September 2001. What I think we have seen since in some quarters, and always need to welcome in profusion, is the rebirth of curiosity—a motivating force that comes from the sense that we do not have all the answers and that, in fact, what moves us along into the future are the questions we pose. Curiosity is certainly a faculty we share across species boundaries, and if fueled by the energy withdrawn from irony and the wisdom culled from earnestness, it could lead us to break through some epistemological impasses.

Our Animals, Our Selves, Our Stories

Science tells stories—scientific stories, reason brought to us on a shiny platter. And what we see on display is the meager diet we have come to subsist on in the face of fewer and fewer stories that might capture our imaginations. The wolf in sheep's clothing. The monster in the mirror. The ape in the human. But isn't there more and more human in the ape than has been acknowledged in past centuries?

The men and women in primatology have gone about producing knowledge that had not been previously on view—some of which I offer for your consideration.

Scholars have been forced to think far beyond their fields as we have come to understand how no one "science"—natural, social, or human—can fully answer the questions that nag at us as we go about sorting through data and stories we don't know how to categorize. Our transgenic creatures—donkey–zebra hybrids, goat–sheep chimeras—embody something akin to the gargoyles of our millennial turn. Now we have the technology to track through DNA new origin stories: "mitochondrial" Adams and Eves who have left their genetic imprints on all of us. Science examines its own interpretive premises; social sciences insist on method and rigor. The very method and form those stories take is what has often distinguished what some call the hard truths from the softer.

On the evening news in May 2001 (what a prelapsarian moment that seems) there were reports of a "monkey-man" loose on the streets of New Delhi. Reporters there were interviewing, metropolitan-style, anyone who might tell about the attacks of this creature who had left some people badly scratched and bitten. The voice-over continued, and the camera retreated, pulling back to show us the sight of monkeys of various shapes and sizes running, at a clip, across a busy thoroughfare—pedestrians, like the humans, though their knuckle-joints as well as their feet touch the ground. The journalistic voice of curiosity and condescension informed us that monkeys are sacred here—like the cows we've long known about. This is a new kind of urban legend, a narrative eruption that speaks to fears evoked by the specter of where our animal-like and animalistic selves roost. Some struggle over how to cohabit exurban and suburban space with deer and bears; some hunt their cousin primates for food, decimating a population of apes, while others work to defend them by writing a Declaration on Great Apes.

As human primates, we do appear to be less insulted in the twenty-first century about our resemblance to our ape kin than seemed the case for observers at the end of the nineteenth century and the beginning of the twentieth. This is not to say, however, that we aren't still somewhat troubled when faced with differences that confound our sense of uniqueness and its usual translation into ascendance in the sphere of species. How we come to learn from the nonhuman animal realm, and what we may surmise as a result of all our research there, is simultaneously privileged and provisional, its proper uses yet to be determined. And this is true whether we are studying animals for medical or sociological insights they might provide to help us live longer, better, more scrutinized lives.

As I write, there are apes in the wild and in captivity, in zoos and in labs, in forests and in circuses. Some of the very fortunate are finding their way to retirement communities such as Chimp Haven, described by Charles Siebert in "Planet of the Retired Apes." And as always they live in our imaginations. Those who study them also carry on imaginary relations with them; Robert M. Sapolsky's autobiography is called *A Primate's Memoir*. The eminent Dutch primatologist Jan van Hooff, de Waal's teacher, speaks retrospectively about his early attraction to studying the furry and the scaly, not the hairy beasts among whom he came to work. Such imaginary relations constitute a crucial aspect of how deeply we recognize our kinship with our nonhuman others. De Waal talks about the "spark across the species barrier" (*Bonobo*, 1), which is the experience of recognition (sometimes love) that we associate with mutual eye contact—whether between us and our household partners we call pets or in a glance that might occur during a quiet moment at the zoo. These are the moments—some daily, some unprecedented—that call us into relations of obligation and care as the "centerpiece" of human–animal, ape-to-ape interrelations. Whether we kiss or kill, we see that kinship is a site of merging and mixing it up. Inter-subjectivities are sites of struggle through alterity toward what some think of as universality; the balance between the two ways of coming to terms with conflict and connection is in question; it's in motion. This young century asks for as thoroughgoing an understanding of what we may have in common as the last demanded we face up to our differences. De Waal asks that we consider empathy as a starting point, not an end. Where might we be able to go with a shifting dialectic between the particular and the universal?

Conversations across the species barrier have a kind of periodicity—they keep coming around. There is a photograph taken by Weegee, an urban photojournalist of Depression-era America, that captures an earlier moment when the culture was struggling with issues of who belonged in the social network of care. The photograph is of "Sherry Britton, Showgirl," reading a book while waiting in the wings; she is dressed and ready to go onstage and, no doubt, literally kick up her heels. The book she is reading is *Apes, Men, and Morons*, its author's name, Earnest A. Hooton, plainly visible on its cover. Hooton was a luminary in the physical anthropology department of Harvard University in its early years. Discarded and disregarded because of its unholy alliance with eugenics, this earlier seduction across the species barrier is rather lost to history. An authority on primates at this time, his is the whimsical voice behind the discourse of the 1930s that would attempt, post–Monkey Trial, to put in perspective the current research as it revealed ourselves to us through what we believed to be true of our ape cousins back then. Today the word *morons* of Hooton's title is discredited

"Sherry Britton, Showgirl."
(Photograph by Weegee [Arthur Fellig]. Hulton Archive/Getty Images)

as an indicator of any level of intelligence. Nevertheless, it still roams the halls of sophomoric humor to indicate unacceptable behavior, a general term of opprobrium for one deemed stupid. In Hooton's day there was an effort, as there is now, to bring the biological to bear on the social. And there was also a climate where science and religion were not at ease with each other's forms of discourse and evidence, and faith aimed to override the knowledge emerging rapidly from many of the still new social sciences. In his introductory remarks, Hooton cautions and assures his readers that science and faith will find themselves allied in his secular work. It does not seem likely that genetics—as prefix, suffix, modifier, and substantive—will retain its associations only with the nightmarish, the grotesque, and the ridiculous, as eugenics came to do. But it also is clear that once again we are living a cultural conflict about how we designate ourselves in relation to our next of kin and kind.

I want to explore two moments in reading; the first is in the realm of nonfiction. Currently a member of that same department at Harvard, Richard Wrangham is the coauthor with Dale Peterson of *Demonic Males: Apes and the Origins of Human Violence* (1996). Reading Wrangham and Peterson's book on masculinity and aggression ("demonic males") among the great apes, I sensed that I was listening in on a new master narrative of gender being sketched out in the remaining primeval forests left to us to survey. Here was a new "Congo" of the late-twentieth-century mind told by way of interpretations of the behaviors of those now designated our nearest DNA-kin—the bonobos and the chimpanzees. Wrangham and Peterson's focus in their penultimate chapter ("Taming the Demon") is on the stunning fact that bonobos are alone among the great apes in not banding together in groups of males with the specific aim to harm another of their own kind. Frans de Waal has become the go-to primatologist to bring this research into public culture, with his most recent books: *The Ape and the Sushi Master, Our Inner Ape*, and *Primates and Philosophers*. One of de Waal's earliest successes, *Chimpanzee Politics*, was given as a primer to some of the newest members of the 1994 midterm United States Congress by their theorist and ideologue, Newt Gingrich, as a way of learning from our animal brethren their strategies for power.

Among the bonobos, as Wrangham and Peterson explain, we observe the habits of a fission–fusion culture where groups and individuals separate for parts of the day or season and develop rituals around how to come and go from and with each other. When assertion of status is the order of the moment, for most primates these are occasions for displays of power and position; however, for the bonobos these moments are lubricated not exclusively by aggressive maneuvers (such as kicking, throwing, chest-thumping) but by what the scientists who study them call, in shorthand, g-g rubbing: genital-to-genital contact, initiated and repeated, across intraspecies differences—of age, sex, or kin group. The bonobos appear to be organized by what some experts call female dominance and others call codominance. How is it that chimpanzees and bonobos offer us such divergent strategies in their conflict-resolution methods? How haphazard and variable can such practices be? Wrangham and Peterson, citing Charlotte Perkins Gilman, turn-of-the-last-century feminist utopian thinker, pose the question of whether the behavior found among the bonobos is to be found anywhere among humans. They continue: "What hope then for taming the demon? . . . Envisioning female power. In her 1915 novel, *Herland*, Charlotte Perkins Gilman tackled the problem of the demonic male by considering what *Homo sapiens* could become in an imaginary world miraculously freed from the constraints of the male temperament and male-dominated political systems" (*Demonic Males*, 236–37). Citing Gilman herself, they then note that "with the loss of men came the loss of fear" (237). How do exceptions—such as the female-dominant or male-and-female-codominant bonobo culture—rewrite the rules of ape culture? The rules about power and desire? How might we gather ideas about how the world could look new? Why not look at worlds that are coeval, contemporaneous with our own? Why not invoke them to remind us of possible pasts as we map the futures we hope to have? We might wonder what other forces could constitute our relations across and through differences? What might we glean of our resemblances to and differences from both the chimpanzees and the bonobos? They and we share a history, as well as a present and a future. They are not simply some atavistic cultural poor relations. Though dwindling in number in the wild, their "captive" populations are still there for us to visit, to study, and to learn from.

Thanks to a Congo dialect, we have named our next of kin after ourselves; the word *chimpanzee* means those creatures who mock man, the human. They imitate us, as the bestiaries of old pointed out. The us they imitate are humans, more specifically, men. When chimps return from foraging, they reestablish their group relations through displays or performances of aggression. Chimpanzees' closest kin in the equatorial forest are the bonobos. Though they were long taken to be the same but simply smaller chimps, we have learned more recently how they also share traits and habitats (south of the Congo River) with gorillas as well. It is no accident that scientists were able to see them anew sometime in the 1970s, for it is at this particular time in history that both feminism and what was called sociobiology were challenging disciplinary borders. What was observed, first by a Japanese research group and later by an American one, both at research stations in the wild, was that members of these groups stabilized their interrelations through displays or performances of sex. Not necessarily heterosex. Not necessarily monogamous sex. Not even necessarily coupled sex. Genital-to-genital contact was observed to be taking place between and among all members of the group except, according to some, mothers and their male offspring.

I am put in mind, not surprisingly, of Margret Rey and H.A. Rey's *Curious George* books—the constant protagonist a monkey cousin to us "greater apes." Curiosity is what gets George inevitably into trouble, but it is also what makes him heroic. He is a problem child and a problem solver. That his curiosity is awakened by being abducted from his home in Africa by the "man in the yellow hat" and brought to New York City makes him a particularly apt choice here. The place of imitation is a curious one—like George. Consider the countless times you have observed humans and their kissing cousins mirroring each other during a visit to the zoo. We cannot avoid remarking the resemblances between us and them. The value we assign this resemblance tells us something of what we need to know about how we think about the differences between species. In August 2005 the London Zoo had on display for a few days, and with the appropriate identifying tags, a group of eight humans on the other side of the bars from the visitors, and next to their great ape cousins: a cautionary tale, a historical update, and an opportunity to consider species life. We take refuge and consolation in the reductive ideology of human as animal—bestial, instinctual, primitive. What the primatologists teach us is that our hairy ape cousins know not only how to pursue conflict but also how to reconcile and repair their connections. And certainly we have much more to learn from them than they do from us, since we know much more about our curiosity regarding them than vice versa.

Narratives that emerge from our ever-shifting scientific knowledge come to bear on how we live in the brave new worlds that are not fictions. We must learn to scrutinize carefully how the reading of our distant pasts underwrites our itineraries into not-so-distant futures. I said I wanted to explore two moments in reading; the second is to be found in fiction. Will Self's novel *Great Apes* (1997) takes the Nietzschean call for becoming-animal about as far as I have seen it go in contemporary literary culture. At the Great Ape Project Web site (greatapeproject.org) we are greeted by photos of the four nonhuman members of the order of primates, our next of kin: orangutan, gorilla, chimpanzee, and bonobo. Many readers are surely familiar with the image that might have graced a twentieth-century science bulletin board or museum display: the figures of the great apes as they morph from larger, taller, stronger, and even gracile, but not quite vertical, to the fully erect, thinner, less hairy, and graceful human specimen. Self's novel takes that iconic evolutionary chart and spins out a tale with the following conceit: chimps, not humans, are the primates inhabiting an otherwise completely recogniz-able contemporary London; orangutans and gorillas do not figure here.

As for bonobos, they represent the economically disen-franchised minority whom we catch glimpses of passing their time on street corners, drinking, smoking marijua-na, and leading generally marginal lives. They are "raced" and "classed," so to speak, in their indolence and lack of employment in the project of citizenship. Self's bonobos are not Wrangham and Peterson's except in their lack of demonic aspect; and they are not de Waal's either in their disconnection from maternal elders. We might understand Self as offering up a representation of what was understood when bonobos were first named as apart from but related to chimps—they were called pygmy chimps because their skulls were first noticed to be smaller than those of chimpanzees. The matter of size was rendered a matter of value as reflected in their naming by humans. Making them less socially, politically, and economically powerful than chimps is the gambit of *Great Apes*; nearly coincident with Self's novel is the appearance in public culture of the news from among the primatologists that this was a serious misnaming.

Self's novel raises the power issues and places them in the foreground for our pleasure and provocation as we learn to use metaphors about so-called alpha behaviors in our self-understandings. Self's inversion of human and chimpanzee is complete down to his use of chimp slang and chimp versions of ethnocentrism, racism, sexism, ageism—the stock prejudices we have all become aware of in recent decades. Mirrored back to us by this text are the ways that the naming of differences prevails in this scenario. Like humans, these chimps use divisions of labor to express power, and lines of force are certainly gendered.

The novel's protagonist, Zack Busner, is a renowned psychiatrist whose professional turf is being challenged by a younger colleague. One of his illustrious patients is the artist Simon Dykes, who is having a psychotic breakdown, marked by his thinking himself to be human. A scandal will end Busner's career, and the once respected doctor will be driven into retirement by his peers, who behave like ape-men in groups when power is up for grabs. Self has resituated Wrangham and Peterson and de Waal, remaking science in the matter and manner of fiction.

We can look at literature and zoology for examples of how we mirror our near ones in the species world—our conspecifics. But in this mirroring we may long to know what they see when they look at us. Self's novel is an effort to depict such a reversal of vision, and to create this vision, he casts humanity as delusional and in a psychotic state of rupture. When Simon Dykes in his nightmare of being human freaks out at the sight of the uncovered hairy legs of the chimps attending to him at Charing Cross Hospital, one of the doctors proposes they create something like "trousers" before entering his room so as to calm his responses to their examinations and questions. One of the sure signs of his mental illness is the fact that he won't engage in grooming. Characters in this novel are grooming each other constantly and in a manner to express their intimacy with each other. How touch occurs, how much touch occurs, and who initiates touch reveal relationships. Mating, another form of frequent skin-to-skin contact, is random, frequent, and predictable any time a female is in estrus. Bottoms are a site for fashion statements: "swelling protectors" come mass-produced and in designer versions. Simon's horror of touch, his insisting on sitting vertically, his refusal to make a nest are all symptoms of his having lost his place among his kind. And his breakdown occurs the same week as his gallery opening. All hip London is abuzz; the remedy is to bring in the famous Zack Busner, radical psychoanalyst, marking the seriousness and celebrity status occupied by Simon.

It is always humbling to see oneself through the eyes of another—sometimes also infuriating—and it can produce responses that can turn deadly. But to see humans through the eyes of these chimps who are our contemporaries, our coevals, in every way but how they look, how they move, and how and when they mate is both humbling and comic. We are familiars to the chimps of London, familiar yet grotesque. We are in their bestiaries: "the human with its ghastly exposed skin; its repugnantly bulging hind parts" (Self, *Great Apes*, 186). The film circulating in London at the time of the novel is *Planet of the Humans*, so thoroughly does Self spin his conceit. The power struggle among the main characters, however, is where the gambit of the novel lies. For it is an issue of succession among senior males for dominance in the profession of scientific research that sets the story in motion. The artist and the scientist vie with each other over the question of sanity or lack thereof.

Self's novel offers a narrative petri dish where we can also read the synthesis of power politics after Foucault and feminism. Look, for example, at the following:

Sarah squatted back in her seat and scanned the copy of *Cosmopolitan* she held open with the toes of one foot, while turning the pages with the toes of another. Advertisements for artificial swellings, swelling-enhancing clothing, swelling clinics, classes and manuals on how to get the best out of your swelling. And the confessional pieces: "I Consorted with a Male for a Year!," "I Joined Three New Groups in One Oestrus" and so

on, and so on. Mating, mating, mating, Sarah thought
to herself—that's all these female magazines are about,
as if it were the only thing that mattered. (130)

Self's novelistic device that grants chimps words, permit-
ting dialogue, all the while reminds us not that he or she
"said," but, for example, there is Sarah turning pages with
her toes, or Busner signing "with chimpunity." What we
learn from the primatologists' tales is how they translate
across wordlessness into what can only be called "languages." Frans de Waal, in a recent book of photos, *My Family Album*, with minimal text, teaches readers about how
grooming practices are cultural and vary widely among
apes from group to group, whether captive or in the field.
In Self's novel it is also the time of genetic breakthroughs
that arouse the evolutionary beasts to stalk us again, con-
ceptually speaking. The bestial in the human is upended
at the close of the twentieth century; we ask now about
the human in the bestial. And for that shift in perspective,
Self offers up a literary case study in our efforts to negotiate our place in the kinship network that would take more
"seriously" our cousins, the hairy apes.

Great Apes is a tendentious text—a thesis text—a political novel. What is "greatness"? What is "apeness"? Except
for their hair, their chimp-speak dialect, and their public
displays of sexuality, the apes in question are everything
we seem to mean by *human*: they struggle for power, they
quest for pleasure, they find and make work in the world,
and they try to keep their near and dear ones in proximity
until subadulthood has passed and a proper consortship
has been made. What Self does in bringing science into

literature Wrangham and Peterson do in bringing science to public culture. Different arguments may be made, but both books are involved in efforts to inquire after the primate's relation to aggression, dominance, hierarchy, and the violence these may do to life among intimates and strangers. Self's chimps may have mastered the bloodiness of chimpanzee politics, but their aggression, when sparked, may turn scatological as they "spray" their feces when anxious, aroused, or cornered, as Simon does when apes try to help him in the hospital. A strategy of primates, Wrangham and Peterson point out, is to "dominate the other group. Remove them, perhaps. But once they give up, let them go. Don't try to kill them. Most primates are satisfied with seeing the rear end of their opponents" (*Demonic Males*, 131). It is chimpanzees and humans who differ from other animals in searching out deadly violence "deliberately." We might imagine Wrangham, Peterson, and Self in a room together, telling their expert tales of murder and carnage—the spray of aggression and the mess it leaves, whether shit or blood. Where Self stages a full-dress imagining of an inverted world, Wrangham and Peterson draw the curtain back on various scenarios of what anthropologists and primatologists have witnessed and reported. But all three wish us to spend our time with them, learning about our demon-cousins so that we might be able to study more closely our own aggressive behaviors, urges, instincts, and histories and resituate our efforts at reaching across these boundaries that are usually so carefully patrolled.

Fiction, fact, and science: human–ape chimeras just waiting in the wings to make their entrance into our universe somewhere in the chain of being. We are in a strug-gle, negotiating the animal and the genetic diasporas to see what will be borne out of them and how we will take account of and care for the ones we are bent on creating, whether through our ever more external bodyminds or through old-fashioned methods of bearing and rearing our kin, our kind, our flesh, our bone, and our blood.

. . . Them Bones . . .

The great ape we observe at the genesis of *2001: A Space Odyssey* hurls a femur into the air that returns into the narrative as a spaceship hurtling into the future—one now already past. Just like the cartographers of the pre-Columbian period, who had their sea monsters lurking at the edges of the known world, we are rereading para-dise at a mitochondrial plane—there are seven daughters of Eve, we're told, by Bryan Sykes, a geneticist in Britain whose Web site (oxfordancestors.com) invites you to sub-mit a saliva specimen and find out your DNA-mother. The Y-chromosome kit for the sons of Adam is now also available; the data-crunching comes quicker than we can absorb; Sykes's *Adam's Curse* tells the story of these more recent discoveries. The biopolitics of everyday life are ask-ing that we become more knowing about those with whom we share characteristics we generally class as animal—not human. And so are the postmodern fables for the twenty-first century, as we rewrite the care of the self, for us and our others.

Kinship is the realm of care, as in care and feeding. It is not news that wherever survival depends on it, our human species may cooperate with another. We may take up the freedom and with it the responsibility to care for those across the species barrier. Domestication brings the love of animals, admittedly long after the symbiosis of habitat brings familiarity. What is the current state of affairs across the differences between human and nonhuman primates—our ape kin? Those ape cousins—how formal or informal are our relations? Whom may we or may we not kiss, relationally speaking? And to move back into the human–animal realm, aren't the hinges of kinship found in the questions that swirl around whom we kiss and whom we kill?

7 AGAIN, A DECLARATION OF RIGHTS

There is a document circulating on the Internet that we might expect to find in Pierre Boulle's novel *Planet of the Apes* (1963) or Will Self's *Great Apes*. But it is to be found in the more virtual realm of academia—drafted by ethicists, philosophers, scientists. It is called the Declaration on Great Apes, echoing the documents of an earlier time declaring the rights of man and woman. It speaks of the great apes as our "disquieting doubles" and makes the case that they be included in an "equality of community" in this age of globalization. Even as we grow into this new century, with more ubiquitous surveillance and suspicion evoked daily, we also find that those thinkers who engage possible futures enjoin us to open the portals between and among strangers so as not to fall into solipsism, to draw in our borders, to re-make our caves. The self must be kept intact so as to enable contact, so as to summon the permeability that is, to my mind, a source of radical hope. And this possible porous-ness must extend to those strangers and neighbors, whether near in time and space or distant. Two phrases that appear as slogans in these times express oppositional views of how we might see our same, self, others: there is the closing of the mind and self of "not in my backyard," also known by its acronym, *nimby*; and there is the sign I often see in my neighborhood that appeared after the war in Afghanistan and more so after Iraq: "not in our name." One *not* refuses and singularizes, as seen in its first-person form; the other invites collectivity and announces it in its first-person plural. The cave as metaphor and habitat tends to keep groups small and defensive; the house, with its doors and windows, at least provides for openness to strangers.

The call of this Declaration to remake the lives that must matter is a practice that Walter Benjamin's angel of history would herald as a historical epistemological shift—where we, as humans, demand for others whom we deem unable to speak for themselves that to which they are entitled. What rules and regulations would regiment our seeming need to classify and order humans in groups? What would constitute ideas of the heroic and empathic in the living of everyday life? These are speculative moves that open the space to conceive the future and perhaps to shape aspects of it. In addressing the future, which we have no choice but to face, perhaps we would be helped to recall as our talisman Benjamin's angel of history, who is propelled into the future but whose glance is back toward the past.

We human primates attempt to structure forms of living that are greater than what Giorgio Agamben refers to in *Homo Sacer* as "bare life." The declaration of their rights—those of the great apes—may, however, also entail new and more intimate ways of enslaving their capacities for reciprocal attachment. In fact, in the Declaration on Great Apes the first point made is their "right to life," where we learn that those "in the community of equals may not be killed except in very strictly defined circumstances, for example, self-defense." The critique of rights discourse has come most sharply from within critical race theory and feminist legal and political theories. How and what have the framers of the Declaration on Great Apes learned from earlier framers of the rights of others? Frans de Waal challenges some of the thinkers who declare the rights of animals, especially the great apes. In *Primates and Philosophers* he asks, provocatively, whether having granted apes their rights, we would wish next in our American fashion to expect they will need lawyers, too. Peter Singer, one of the framers of the Declaration, responds to de Waal that we might usefully take apes, our proximate kin, our cousin species, as models for how we care for and are obliged toward all animals in our coexistence.

If kinship is to be reckoned in terms of care and love, then we have to come to terms with the effects of the fact that those we designate "lovers," those who love others outside the bounds of their own kind, are often subject to the greatest opprobrium; and, of course, we also designate as "lovers" those whom we take to be our intimates. To accept and learn from our next of kin across species barriers returns us to how we care for and love those more proximate, those whose genes, genealogies, and names are passed along across generations. Wouldn't it then become imperative to disentangle the moral from the social and recalibrate the cultural and the natural? It is as humans that we have marked our worldly manifestations by the cultural imprints on the natural—what we make of that which is given. The second half of the last century removed any sense that the utopian might be reconstructed other than as fantasy, and often as fantasies that revealed their dystopian dialectics quite clearly. Utopia was sometimes the good (*eu*) place, sometimes the no (*ou*) place; now it is neither. The shift might be more advisedly read as an *elsewhere* we might get to map and inhabit if we can learn, through mutual curiosity, as much as we can about the past we take into our futures. The call for a "community of equals" defines one of the aspects of this new century and what it has to offer as we look back into our deepest pasts in order to prefigure and configure our necessary futures.

The Italian philosopher Giorgio Agamben speaks about "bare life" as what we have come to know since the concentration camps, where the sacred—in ruins—was rewritten for the latter half of the last century. The sacred subject, according to Agamben, has long been one who may be murdered—the exterminatable, we might say. But though the sacred subject may be murdered, the murderers do not take themselves to be guilty of homicide. Their others are rendered bestial—made from fables of wolfmen and transformed into real live Jews, Gypsies, queers, and others in the list of those who would be liquidated as classes of humans deemed unnecessary to care for. In the rewriting of life after the Shoah, one of the prominent voices was that of Emmanuel Lévinas, the philosopher of otherness, of alterity. His thought demanded in the encounter between others a near erasure of the self, an erasure that could even extend to the victim's being asked to empathize with the perpetrator in their mutual (in)humanity. This impasse has engendered a more recent critique that restores the self, and the same, to our understanding of encounters among strangers, others; the foremost voice of this critique and rewriting of the ethical against Lévinas for this new century is Alain Badiou. This bloody boundary between selves and others is a site that requires our close attention. Blood shed. Blood mixed. To make a new history. Possible futures. The biopolitics of everyday life ask that we become more knowing about those with whom we share characteristics we generally class as animal—not human. Reading the fictions, films, and postmodern fables of the early twenty-first century, we engage in rewriting the care of the self, for our selves and our others.

The Great Ape Project asserts that it is not ethical to wait until all humans have been liberated to demand the rights of our ape kindred. That their torture should be prohibited—one of the objects of the Declaration—in fact, ought to elucidate just how far we are from the dissemination of rights of freedom among our own kind. In an essay intended to explain further implications of the project, Paola Cavalieri and Peter Singer address the concept of the "person" as a status to be considered for those of the great apes whose rights are being claimed. Cavalieri and Singer's work comes at the intersection of the growing fields of applied ethics and primatology in the 1990s. Returning to thinkers as ancient as Boethius and Aquinas, they ask that readers recognize a capacity for reasoning among our ape cousins, as repeatedly observed by those working with apes in the field and in captivity. They have biographical lives, not merely biological ones; they have culture and are not simply or in some complex way relegated to nature. Their use of language, long maintained as that which separates us from animals, is basic to many kinds of research going on with apes in captivity. Their ability to teach what they have learned to others of their own kin(d) further distinguishes their place in the realm of what has historically been signaled by the category of "personhood." De Waal points out the rather recent knowledge gleaned from genetic researchers that indicates a capacity for sociality among bonobos that is absent among chimpanzees; this certainly adds to the mirroring we find there and to their more peaceable, less murderous, natures and cultures.

To make this point, Cavalieri and Singer retell one simple, stunning story. Geza Taleki, a longtime observer of chimps in the field, tells of an evening when he was watching the sunset over a ridge and saw two adult males coming from opposite directions, face to face. They meet, they shake hands and sit down, and they proceed to watch the sun set, as their human observer is also doing. The literature on chimps and bonobos is replete with stories as compelling as this one. While some may read Taleki's moment as exemplary of the imperial gaze long nurtured in Africa, it is the kind of narrative that also enables a notion of a community of equals who deserve to be protected from torture in the laboratories where large populations have for generations spent time in cages. From here it is not a huge leap to the analogy of slavery—its long existence and the protracted struggle for abolition—also based in Western notions of personhood. These animal others are also persons in need of guardians who will act out of a sense of ethical, empathic obligation.

Newly Born Chimpanzee!

Providing a concrete case of what the philosophers demand is the latest news from the retirement community of

Chimp Haven. In this "planet of the retired apes," described by Charles Siebert in the summer of 2005, are chimps out of cages, living on islands in Louisiana—neighbors, as it were. And 2007 brings news of a birth in the retirement community: on January 8 of that year, Teresa, age forty, gives birth to Tracy. And here's what puzzles their human guardians: Who is the "father," given that the males were vasectomized before their move to the island? So paternity tests are being performed. And I can't help but wonder, Then what? Will there be some effort to force Jimmo (in his forties), Conan (twenty-one), or seventeen-year-old Magnum to provide child support of the primate kind? I also feel a kinship with Teresa, a forty-year-old single mother—a cousinship. And for the primatologists who study patterns of child rearing among chimps, what will emerge from the efforts of the group to participate in their care for their newest young one? While not in cages any longer, or in circuses, or in chains, these great apes have undergone nonelective surgery (obviously failed, at that) because the humans who guard and protect them thought it for their own demographic and social good, no doubt.

The determination of paternity in this case will lead to another surgery—those are the plans of those who oversee the islands that make up the community. In any event, the very idea of a "retirement" community has been turned on its head with this birth announcement. Through the efforts of a local philanthropist, we learn that Teresa and Tracy will not be separated—why would such an option even be considered given all we know about chimp cultures? If we are arguing for their personhood, as the framers of the Great Ape Project would have us do, how would the separation of newborn from mother figure in our guardianship of these members in the "community of equals"? Yet this kind of severance among kin was common zoo practice until recent decades.

Additionally, how to read the urge to determine paternity? Our urge, not theirs. Who'll be watching out for the behaviors that so shocked the primatology community in earlier decades of males committing infanticide so as to send receptive females into estrus again, thereby allowing them the opportunity to mate again, ensuring their genetic survival and place in the social hierarchy? We are told that Teresa was missing one morning and reappeared the next day with a newborn. Surely those present are watching to see which females and males will cooperate in the rearing of this newest member of their community of equals—a newborn in the retirement community thereby reconstituting and redefining the community by its very presence.

I, for one, am poised for the stories about who joins in the community of care. How far do the kissing cousins extend? What is their range? Who will be trusted to kiss, not kill, this young vulnerable one? Where will the negotiation of those porous zones draw out cooperation and conflict? And what does this still studied space—yes, without bars—teach us about the culture they make in our company and that we make in theirs?

When Charles Siebert wrote about Chimp Haven in 2005, it was about to be opened to its first residents. He described it as a "monkey Delray Beach," mixing the primate metaphors, but the point of the comparison was the human notion of what retirement looks like: age-determined communities. And Amy Fultz, one of the resident behaviorists, explained that the vasectomizing of the males would allow them to "express their normal behavioral repertoire" (quoted in Siebert, "Planet of the Retired Apes," 31). Well, they have. And this late-life newborn for Teresa, who has been a mother eleven times before but not in thirteen years, gives her and her guardians the occasion to see what relations can be established under these newly "ideal" conditions. How will she and her conspecifics, both human and animal, arrive at ongoing relations of care for the most vulnerable among them? Where will chimp child-rearing practices become interpreted and observed and enabled? How will those of us who care to learn watch, study, and imitate them? Aping their versions of "personhood"?

Given that we are living in an era when *personhood* seems not so readily applied to those who occupy the same side of the species divide—mustn't we make the case more wholeheartedly, and more honestly, for opening up the definitions?

8 FROM CAGE TO CAVES

As the leading metaphor of kinship through this survey of stories, cousinship is used to describe the relationship between great apes and humans. Not a quantitative sort of critic, I haven't even tried to keep count of the occurrences of the metaphor, so let me cite just one of the more recent spins on our closeness to the great apes: just weeks ago I was passing through the National Museum of Natural History, in Washington, D.C., where the great ape exhibit enjoins the public to "Come visit your relatives!"

It would appear that our appetite for seeing ourselves in them, seeing them, seeing them seeing us is bottomless. The discovery of what neuroscientists have named mirror neurons leads scientists to speculate about the origins of language and might also be invoked in the ongoing discussion of empathy here; these neurons fire when a subject not only performs a certain task but also observes one of its conspecifics performing the same task. I'd speculate that these mirroring moments between animals and humans lay bare our sense of kinship, our sense of cousinship with these same-self others. From the behavioral sciences and from philosophy, we encounter thinkers asking about the preeminence of the face and the eyes in the empathic encounter. Face to face, eye to eye—such proximity is required for cooperation and also underwrites conflict. This kind of mirroring and the curiosity it evokes is an exhilarating and open way to approach the questions I am trying to address. I have also insisted that kinship, cousinship, doesn't pertain alone to those we "kiss" but also to those we "kill" with our curiosity.

In the arena of films that address the human–ape contact zones, there are many to choose from; many play on the kissing side of the cousin relation. But one of the more recent films that draws exclusively on our shared killing nature is a postapocalyptic scenario set in London called *28 Days Later*. An opening scene has several animal rights activists breaking into the Cambridge Primate Research Center, planning to release the lab chimps. We see one particular chimp splayed on a table, arms and legs in restraints. Before the intruders can begin to open the cages, a scientist appears and pleads with them not to follow through with their plans. He explains that the chimps have been infected—with rage, the rage of humans. Having released the chimps from their cages, the activists become the first victims of the rage virus. The screen goes dark and reopens with the words "28 days later" in the lower third of the screen. On a gurney, positioned like the chimp in the lab, is a human awakening from some long-acting anesthetic; tearing at the tubes and bandages on his body, he frees himself and begins to try to find out where he is. We see London abandoned and devastated by violence, signs of which are omnipresent. After a long while of calling out to anyone who might hear him, he comes on two strangers who explain that the virus has spread swiftly with catastrophic global consequences. We are in full dystopia in 2002, the release date of this film. This is no gentle thought experiment like the *Planet of the Apes* series from the late 1960s; it is no cozy comedy like *Human Nature* from 2001. Rather, in this scenario we share the vulnerabilities of our closest animal kin, and like them, when enraged with the virus that humans have used to study and infect them, we too become capable of spontaneous murder.

A story of trauma inflicted in the space of twenty-eight days—a month, a cycle of the moon—and what there was of "humanity" is reduced to its handful of representatives: a young white man, Jim (Cillian Murphy), and a young black woman, Selena (Naomie Harris), are looking for but also fearing potential other survivors when they find Frank (Brendan Gleeson), a white father, and his daughter, Hannah (Megan Burns). These two are signaling to any others out there by the flashing of Christmas lights from their apartment, the only one that appears to be inhabited in an urban high-rise building. Now a rapidly constituted foursome, they are forced to fend for their continued survival. Their ethical decisions are instantaneous rather than deliberative. Their small number as survivors nevertheless constitutes them as a band who will become kin, related, interrelated as they ensure both their own and each other's survival for as long as possible in this devastated landscape.

In *28 Days Later*, Selena is figured from the start both as the wise woman and as the heartless survivor when she explains to Jim that she will defend her own life first—her decision will be made in that proverbial "heartbeat." Selena insists that "staying alive is all there is." On a battery-powered radio. the father, Frank, has captured a recording of a man's voice that claims to come from north of Manchester, saying, "The answer to infection is here." The four then flee the devastated urban streets of London, and at their first encounter with the rural. Frank is the first to be stopped in his tracks by the site of four horses, two black, two white, and directing the others' glances toward them, he says, "Look, they're a family." A call to empathy, an act of projection, and the dwindling of hope drive them north, where they do indeed find a small garrison of paramilitary men. While there, a stray drop of blood into Frank's eye leads to his becoming infected, and for the others to stay alive, he must be killed. This demands the kind of thinking embodied in what those who engage in moral reasoning have called "trolley problems"; such situations require, according to Peter Singer, "using their reasoning to override their emotional resistance to the personal violation that [killing] another person involves" (quoted in de Waal, *Primates and Philosophers*. 148).

Now these three survivors turn out to be the hope of this self-styled militia of young men in full warrior mode. Their self-appointed commander tries to persuade the newcomers that if they are *all* to survive, the woman and the girl must lend themselves to the task of satisfying the frustrated and lustful desires of these isolated men, arguing further that "women mean a future." At this suggestion, Selena hatches a plan with Hannah that leads to the murder of some of these men as they assault the women and aim to rape the young girl. In a scene designated "Selena's Savior," Jim comes to Selena's rescue as she is being attacked by one of the last of the predatory young men, now infected. As she is about to take aim at Jim, he pleads quickly that he is not infected. Stopping in midstroke, she drops the stick that is her primitive weapon and raises her arms simultaneously as they engage in a kiss that releases each of their withheld desires for connection.

In this kissing-killing knot, *28 Days Later* offers us a re-writing of the primal scene in this newly depopulated universe. The pubescent Hannah, now fatherless, becomes the child/ward of Jim and Selena, who become instantaneously coupled in the face of radical fear and loss. Though blood and saliva are the means of transmission of infection, as uninfected, these two share a first passionately desperate kiss that is witnessed by Hannah. Thinking this is a battle, given that is all she has been witnessing for the last month, she hurls herself between them. "I thought you were biting," she says. Jim explains that they were "kissing." The very figure of the traumatized child, Hannah had also just seen her father murdered before her eyes.

The "genus" being murdered here is not ethnic or racial; it is the human race, species, genus. The "gens" exists between the "family and the species." The exchange that has been entered into is the human infection of their "cousin" species with their rage—which, when returned to them back across species lines, has further magnified their aggression. Each scenario suggests self-defense as the reason behind the killing of all who are infected. It is the last resort—a defense of homicide in the cause of self-defense. Any veneer of morality has been swept away in this universe; all that remains is the minimal porousness and permeability that allow Jim, Selena, and Hannah to start over, survivors, in making a new kind of family, all adopting each other as intimates.

Yet this film—as dystopian as it gets—insists on its own reversal. It perseveres in a vision of a future. Our three survivors are freeze-framed about to escape and crash through the locked iron gates of the garrison when the screen goes black and a second time the title reads "28 days later." We see Selena playing field nurse to Jim, as in the beginning, in a hospital bed, and when not nursing, she is sewing at a foot-powered machine. Hope is to be found in the winding sheets she is stitching together for that moment that they are awaiting in a pastoral setting, where, perhaps, they have joined those horses. We learn that a plane has been seen overhead, indicating that life had gone on elsewhere. Hannah plays lookout, and when she hears the sound of a plane approaching, all three run to set the sheets across the wide ground where they will send an SOS to the heavens above. It is the word *hello* that calls out to the others alive. The final words of the film: Selena smiles and says to Jim and Hannah, "Do you think they saw us this time?"

This narrative imagines the darkness that humans are capable of inflicting on those they take to be their kin and kind, but it shifts the power that arrogates rights to humans and the field of experiment to those nearest across the animal divide. This experiment was, after all, an effort to see what the power of rage might unleash in the lab chimpanzees. Those running the primate research lab, whose locks were smashed by the animal liberationists in the first scene, were, after all, putting back into the chimps what we, humans, had suppressed—rage. This film poses the question of how, in small numbers, we humans would start over in making our caves and perhaps emerging from them to see who might answer the call for help that is the final scene.

9 TREES OF ORIGIN

On the facing page is a diagram that can be found in "Of Genes and Apes," an essay by Anne E. Pusey in a collection called *Tree of Origin*. This schematic narrative has fascinated me since I first encountered it in my reading in the realm of conflict resolution. A growing interdiscipline, conflict resolution includes animal behaviorists, particularly primatologists, and policy makers, especially those grappling with ethical and legal questions. On this continuum we find professionals engaged in thinking about how to take what we find in the forms of conflict and reconciliation enacted nonverbally (but perhaps quite loudly) among our primate kin and refashion these forms for our world of law and the desire for justice.

To situate this diagram in context, we follow Pusey, who begins by asking, "What kind of social groups did our ancestors live in?" I would add a follow-up question: What kinds of social groups can we continue to live in? Two axes describe the impetus for my chapter: the importance of what we can see with our eyes as it affects the living out of conflict and connection—killing and kissing; and the second issue of how what we see brings us close enough

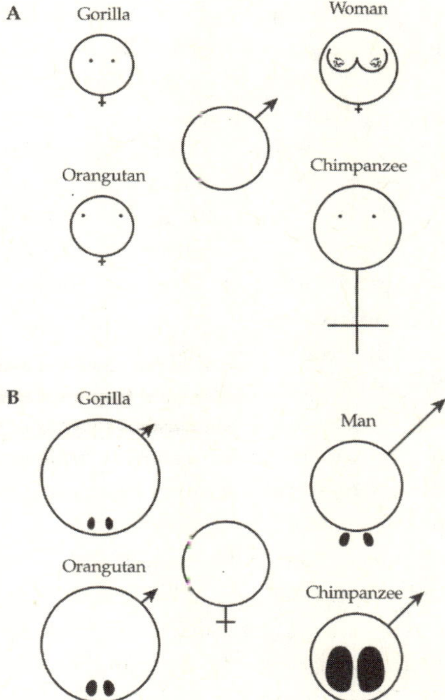

Sexual perception of each sex by the other in four species of great apes: (*A*) the male's view of the female, indicating the degree of sexual dimorphism in body size (circles) and the relative development of the mammary glands and distension of the perineum (crosses) before the first pregnancy; (*B*) the female's view of the male, indicating the degree of sexual dimorphism in body size, the size and position of the testes (ovals), and the relative size of the erect penis (arrows). (From Anne E. Pusey, "Of Genes and Apes," modified from R. V. Short, "Sexual Selection and Its Component Parts, Somatic and Genital Selection, as Illustrated by Man and the Great Apes," *Advances in the Study of Behavior* 9 [1979]: 131–58. Reproduced by permission of Harvard University Press)

for mating, and for murder. These are the two activities where kinship emerges and where it breaks down. Pusey is concerned in "Of Genes and Apes" with research on chimpanzees—their patterns of residence and affiliation, given their group size, which can range from 20 to 120. Pusey says that the fission–fusion culture of great apes is "unusual among primates and mammals in general," yet it does describe us and our cousins: "individuals of the community spend some time alone and frequently join and leave temporary subgroups" (14). This pattern has as much to do with feeding on a daily basis as it does with mating when females are receptive. And the deadly aggression of male territoriality first reported in the 1970s is derived from efforts to defend food sources, and females are extremely aware of and careful about the shifting borders of intergroup hostilities.

Pusey and other primatologists follow the patterns of conflict and cooperation among kin. These relations are becoming more readily traceable from DNA collection through hair, partially digested food, and feces. We learn that cooperation is more likely to occur among those who are close kin rather than those more distantly related. We learn also that incest appears rare and that those who have matured together tend to avoid incestuous coupling—a fact observed and named in 1891 by anthropologist Edvard Westermarck, and named after him as the Westermarck effect. These sets of observations bring us back to the questions of how the "kissing cousins" come to negotiate those who will be protected by the circle of empathy and how the struggle over the dimensions of that circle will remain one of the highly negotiable aspects of life in groups. But kin are not the only deciding factors; neighboring groups of chimps also find themselves tied in relations of cooperation. And sibling relations feature prominently, as well, in the patterns of cooperation and conflict.

Now mating patterns among chimps and bonobos are among the distinguishing aspects that lead to the view that humans are indeed much less promiscuous than our closest relatives. Primatologists are persuaded that among the strategies that determine the multiple matings that occur for females in estrus is the desire to confuse paternity, thereby gaining the protection of themselves and their young by many more of the males in any given group. A term that has come into use among primatologists is *consortship*—a pairing off of two chimps while the female is receptive. It is one of three patterns, all of which ensure her relative safety and the potential for the survival of her young. The two other patterns are mating with as many males as possible and, when their number in any group grows too large, several males cooperating to guard a single female. Jane Goodall and her colleagues were among the field-workers who first observed these patterns. It has remained for others to continue to interpret these behaviors and to sort through some of the evolutionary advantages for groups and the individuals in them. Among the results for kinship and genetic diversity is the fact of many maternal siblings, with paternal siblings rather rare as a result of females' mating with so many males in the primary group or neighboring groups. While alpha, or dominance, behavior matters to the social dynamics of groups, it becomes apparent through genetic testing that in terms of genetic success—that is, reproduction—the hierarchical status of males is nearly irrelevant. Many males "achieve paternity," notes Pusey (33).

When Pusey turns her attention to the comparison with human mating patterns, she notes the universality of human bonds between males and females. We need to recall that those bonds might be between brothers and sisters, as among the Na, or between mothers and sons, as among chimps and bonobos, and not necessarily between the nuclear pair of heterosexual mating. Here Pusey turns our attention to the issue of sexual dimorphism—how males and females of the same species differ in size and attributes. Here is where the patterns among humans in relation to our ape cousins gets especially intriguing.

I offer for your consideration the diagram at the beginning of this chapter: let me simply try to tell what strikes me as both hilarious and worthy of serious contemplation. The diagram sets us in context of our cousin-apes as far as how our attention is drawn to our nearest others, those who call to us through the display of body parts. Sexual display and selection depend on what we see (and smell) and its ability to arouse the bodies in question to meet and mate. Perhaps among the most striking features of human sex selection is the presumptive move from hind parts to front. That would accord with the widespread agreement that we as a species are more likely to be consciously moved by sight than by smell, for example. It is what our walking upright has both given to and taken from our patterns of intimate affiliation. For a long time that very verticality has ensured us our place at the top of the chain of being—remember the apes in the ancient bestiaries with their "revolting hind parts."

But even among our cousins, the display of sex characteristics carries a message across space. If you look at this diagram as a narrative told from a heteronormative perspective of reproductive drives, you will see (1) how men view women and (2) how women view men. The biological symbols for male and female stand in here in their most concretely signed aspects: the circle atop a cross marked female, the arrow at an angle marked male. And here size is certainly what counts. The size of that cross of sexed body and the arrow of penetration indicate just how openly the urge to mate is announced by female sex characteristics and how well endowed males are, based on the circles indicating their scrota and arrows proportional to penis size. We notice that for humans the interest has moved from rear views to frontal: breasts from adolescence on, not simply when lactating, and penis size that dwarfs that of our primate cousins, who advertise more by their scrotal sacs.

The diagram also presumes sophisticated and subtle forms of communication among our fellow primates based on signals of a bodily kind. Here we have the relative size of secondary sex characteristics—breasts/nipples and testicles/penises—as forms that signal receptivity between males and females, and though I have yet to come upon such speculations, perhaps also between males and males and females and females. The chimp specialists do discuss the phenomenon of what they call "penis-fencing" among juveniles who are rehearsing forms of dominance and hierarchy and their positions among their conspecifics.

There are many more maternally related brothers among chimps, whereas females are more distant relations since they tend to migrate to other communities, observing what is called male philopatry—males staying on home ground. And we know from Pusey and others that "cooperative male sibships are prominent in many traditional societies" and usually include the exchange of females in marriage between and among males (34). Pusey takes her questions of analogous or comparable social organization to asking how humans have "forgone high rates of promiscuity" (35). Where primatologists describe chimps as a "party-gang species," Pusey, citing Richard Wrangham, hypothesizes a kind of "bodyguard" function performed by males to guard food sources and against "sexual coercion by other males" (37). In an effort to further explore and understand the development of the pair bond, Pusey assumes the universality of the pair bond, but like most who do so, that pair is usually understood as a reproductive pair and not often as a sibling pair, as we saw among the Na of China, for example. Here, again, my perspective turns to the expandability, the plasticity of forms of kinship, kintimacy—and we look with equal curiosity to our animal cousins as to our human neighbors and strangers. We look to our evolutionary histories and to our contemporaneous contested sites where conflict and reconciliation, proximity and distance play out through our interconnected caves.

The iconic representation of the great apes acknowledges four sets of cultures: gorillas, orangutans, chimpanzees, and bonobos. The last group keeps falling outside the wider circle of empathy. We idealize and sanitize simultaneously what we stand to learn from their conflict resolution methods in a culture where females dominate. Instead, we keep writing stories and scenarios of chimps where the exploration of conflicts and their enactments allow us to imagine a kind of cousinship that is not the kissing kind.

Circling back around to where I began this section, it must be said that I was as surprised to find myself reading primatologists as Richard Wrangham must have been when he turned to Charlotte Perkins Gilman, feminist utopian thinker, in writing about bonobos. For me, it all began with Natalie Angier's article "Bonobo Society." There I and countless others learned about the apes who are not chimps and who resolve conflict through sex rather than aggression. I wanted to know more, and for some ten years now I have been reading, writing, and thinking about these cousins of ours and what they offer as an alternative narrative of kissing and killing. They have been called

"Vernon, male bonobo." (Photo courtesy of Frans de Waal)

the gentle ape; the forgotten ape; the apes that make love, not war. They make a stunning case study of how relations among intimates are rewritten when females are dominant and when the basic kinship unit is mother and son.

Throughout *Bonobo: The Forgotten Ape*, Frans de Waal places a premium on the question of eye contact: between the bonobos themselves, looking directly and seemingly deeply into each other's eyes, and between them and us, as we take each other's measure of the seemingly narrowed space between human and animal individuals. De Waal explains several factors that contributed to the bonobos' becoming perhaps the "last apes" to be counted. This began to change dramatically in the 1970s, the decade of emergent feminism, with the establishment of research stations in the wild, particularly by Japanese primatologists whose epistemological assumptions differed from Euro-American notions of ethology, or animal behavior. In the post–World War II era, ideas about kinship, conflict, and connection were based on the immediate history of our own human capacity for murder after the mass death perpetrated in Europe and Asia.

De Waal is to be applauded not only for his role in public culture as the frequently cited expert about bonobos but also as an excellent historian of this branch of the primatological tree of origin. He is Dutch by birth and obtained his first training at zoos in the Netherlands. He is also the inheritor of the early ethological tradition that gained Dutch researchers recognition in the beginning of the twentieth century—research that emphasized similarities and cooperation rather than the postwar turn to aggression. De Waal looks back at work done in Dutch zoos in the 1930s on differences between chimps and bonobos. A couple working on skeletons rather than live individuals detailed differences only later confirmed in captive and wild populations. Another scientist, Claudia Jordan, wrote a dissertation in 1977 that would later be confirmed among English-speaking researchers about the use of deception among bonobos, a capacity that also suggests an awareness of others' desires, if only to outwit them. The time for discussing bonobo differences needed a more open climate in the discussion of animal sexuality than Americans were ready for until very recently. It must be said that de Waal has insisted in *Bonobo* that we "drag the bonobo out of the obscure corner in which primate specialists have been debating its particularity among themselves" (13). This is my effort to aid in such a project so that we can turn our attention to them in order to see them, and as far as possible allow them to see us seeing them, so that we may more fully explore the depths of these mirrors as they reflect along the corridors of our species-mind.

In the autumn of 1997, I had the unprecedented pleasure of spending time with the keepers of what was then the largest group of bonobos in the United States at the Milwaukee County Zoo. I learned the apes' names, looked into their eyes, saw them with those whose work lives included them. I heard some of their stories from the keepers. Most poignant was Brian's. He had recently come to Milwaukee from Atlanta. He was an adolescent and was not accustomed to being in such a large and wide-roaming group. He was being slowly socialized and integrated into an easy companionship with Linda, the most senior of the females. She had the most calming effect on Brian, who also had such creature comforts as a television. We know rather a lot about the visual culture that bonobos—and chimps—enjoy. We know that those who spent years in labs are either comfortable or anxious around folks in long white coats. We know from Carole Jahme's study *Beauty and the Beasts* that some are particularly fond of color glossies of blonde women, especially those in furs. It should come as no surprise that fur is the tactile medium of bonobo intimacy, given that a great portion of the bonobos' time awake is devoted to mutual grooming behaviors.

Rules don't simply have exceptions. Rules are rewritten through their exceptions. Bonobos represent a cultural set of exceptions similar to those I discussed earlier among the Na of China. Natal groups are based on mother–son proximity, with daughters migrating out to find their place in a new kinship group. Bonobos keep dropping into and out of the story of our kissing cousins—for they surely merit the designation: they appear to welcome all into their circle of empathy through intimate grooming behaviors that bring the modesty of humans into sharp focus. Our resemblances keep getting sliced ever finer; one way we keep our distance has to do with the fact that they mate frequently, in all combinations and numbers, and in public.

In *The Mating Mind*, Geoffrey Miller offers up for our consideration a history of how, until the 1970s, science and culture turned away repeatedly from the second aspect of Darwin's theories—the fact of sexual selection as opposed to the cliché and rallying cry of natural selection. Where sexual selection collides with ideas deemed potentially unassimilable is the way it demonstrates female mate choice—agency among females. And is it any wonder that this hypothesis might keep getting buried in the world of ideas? Robert Sapolsky is among the more popular science writers who bring to our attention the controversial subject of female choice, its emergence in the literature on animals in general and on primates specifically. In *Monkeyluv*, he clarifies for lay readers that the discourse that lends intentionality to animal behavior "is just an expository device agreed upon to keep everyone from falling asleep during conferences" (41). Although this is certainly an important caveat, enough anecdotal evidence in the literature makes clear that our ape cousins have their wants and that they express them plainly. But beyond this: they are capable of recognizing others' desires, even across the species barrier. In *Primates and Philosophers* de Waal tells the tale of Kuni, a female bonobo who captured a bird and then tried to help the bird to fly: she "climbed to the highest point of the highest tree . . . then carefully unfolded its wings and spread them wide open, one wing in each hand, before throwing the bird as hard as she could towards the barrier of the enclosure" (31). The ape knows that the best it can do for its companion animal is to take it up to the highest branches and release it into the air, or in this case of sympathy for the bird's need—different from her own—hurling it skyward. Such stories, and they are numerous, lead researchers to hypothesize what philosophers call a theory of mind among our fellow apes. This capacity to take the perspective of another is among the prized traits of the human, and it is critical for understanding conflict and reconciliation.

Amy Parish, a prominent bonobo researcher working at the San Diego Zoo, began collecting fecal samples for DNA testing in order to trace kinship ties, as well as other information about the bonobo genetic legacy. Among the bonobos who were her research subjects were some who, once they understood her curious tastes, began bringing samples directly to her, facilitating the gathering of her materials. Mind you, while apes are certainly known to throw feces in anger and in stress, they typically avoid this degree of direct contact. Yet their attention to human interests and culture would lead them to overcome a kind of species-inhibition out of care for and cooperation with the human they had come to know.

For all their less frequent resort to aggression to resolve conflict, female bonobos can become dangerous to males if the males attempt to undermine relations between or among females. Parish has confirmed this in her work and is credited with proving that bonobo females prefer their own company under most circumstances. They can maintain their dominance by refusing to cooperate with males, as well as by using less passive-aggressive behaviors. All this to say that while bonobos may be making their presence known in popular culture bit by bit, they are being taken to be more peaceable—the "make love, not war" apes. We welcome the mirror they hold up to us as members of related tribes. Primatologists speak of bonobos and chimps as "charismatic megafauna" with whom we are most likely to identify, unlike the many other members of the natural orders, so to speak. And as for their promiscuous sexuality, we should note that sometimes they, like us, take their couplings in private, hidden from their others. Parish also tells the following story: after the birth of her human child, named after a bonobo Parish had worked with, she made a visit with her infant to the zoo, where one of the resident females, Lana, had also given birth. Lana brought her infant to show off to Parish and made clear her interest in Parish's child. There was a mutual curious and careful inspection of both infants by both mothers.

Sue Savage-Rumbaugh is the primatologist most closely studying language acquisition, use, and teaching among bonobos. She has worked primarily in the laboratory and is currently head of the Great Ape Trust in Des Moines, Iowa. She did, however, decide to spend some time observing bonobos in the wild. One of the stories she returned with is among the most remarkable in the way of cooperation and communication. On "ninety occasions," as reported by Jahme (*Beauty and the Beasts*, 282) and counted by Savage-Rumbaugh, she watched bonobos come down out of their nests, built nightly, in order to head out to find food. On their way, they would set down leaves, and sometimes branches, pointing in the direction they were headed so as to leave signals for the stragglers.

One last anecdote from Jahme, whose *Beauty and the Beasts* focuses particularly on the women primatologists. We learn that Kanzi, made famous by Savage-Rumbaugh's language work with him, will not be allowed to breed, as determined by those who monitor the species diversity desired to avoid inbreeding. Whether Kanzi has been vasectomized is not clear, but he is provided with materials to enhance his own pleasure. These would include "ordinary women's magazines" (284). And, as I mentioned earlier, Kanzi and other chimps and bonobos are excited by images of "blonde, white women, especially if they are dressed in fur coats" (284). Jahme speculates: "This fact tells us a great deal of how these animals perceive us. They feel close to us because they know they are, even if some of us reject their kinship" (284).

By the 1990s, this research brought the bonobos to our attention in the public realm. But it's nearly a decade since I first learned about these cousins of ours, and to most educated readers to whom I mention them, there is still a look of puzzlement, curiosity, and ignorance at their name. And it's not because they know about these apes under their former designation as the pygmy chimpanzee. It is, I would argue, because our knowledge of their social and cultural forms of interaction defies the "chimpanzee politics" so much more readily in evidence in human life. And it doesn't help that their very name keeps shifting; in *The Last Ape*, Japanese field primatologist Takayoshi Kanō suggests wisely that it would help in conservation work on wild populations to refer to them by the names used (*elia*;

pl. *bilia*) among the people who share their habitat and some of the same favorite foods.

De Waal moved from the chimps to their bonobo concestors on the other side of the Congo River to study their different social practice of resolving conflict not primarily through aggression but through forms of sexuality that are what was once called polymorphous. De Waal sums this up by saying that where "the chimpanzee resolves sexual issues with power; the bonobo resolves power issues with sex" (*Bonobo*, 32). These struggles are not absent; rather, they are dominated by females who keep their male offspring with them for life. The "daughter" bonobos are obliged to leave the natal group for a new one in order to make their own bonds for life.

But this emphasis on their sexual activities is just one aspect of their uniqueness. Those who work with them and also with chimps teach us that their "voices," or vocalizations, are the most striking set of differences from chimpanzees. Bonobos' timbre is of a higher pitch, their range of sounds is more expansive, and the frequency of their use of "language" is much greater. They "talk" rather a lot as compared with our mutual cousins, the chimps. And perhaps most striking, they "talk" with their hands—gestures fill in the picture of their sociability. In fact, de Waal notes, social life is of singular importance among bonobos. While they may not share food as regularly as chimps, there is speculation that this has to do with the greater abundance of their food sources. Hence food tends not to be so regularly a source of struggle but rather one of excitation. Some of the most voluble exchanges among members of a group involve the discovery of food and the grooming and g-g rubbing that ensues until food has been consumed and sometimes, yes, shared as well.

Among the tests that determine self-awareness is the one in which a spot is "painted" on an animal's forehead, and a mirror is made available. Both chimps and bonobos, and more recently elephants and dolphins, reveal a level of self-awareness that brings them into our "community of equals." Bonobos not only recognize the fact of something being there that is out of place but also make use of the mirror to self-groom, to look into their own mouths, and, more winningly, to make faces at themselves. De Waal feels himself the object of their gaze when he engages in such research, as the bonobos make him "see the foolishness of science's obsession with classification." "Were they mocking me?" he asks the reader as he had asked himself (*Bonobo*, 34). There is the moment where object of knowledge and subject of knowledge are fully merged and blurred in a way so as to leave us to consider all that we have historically kept most precious in our sense of ourselves at the head of the great chain of being.

I can't help but hear a potentially gendered narrative being derived out of a species distinction. In an interview that de Waal conducted with Savage-Rumbaugh for his book *Bonobo*, he asks her to offer insights on the differences between the bonobos and the chimps she has worked with. She explains: "In anything outside the domain of social communication, involving object manipulation or spatial orientation, the chimpanzee was reliably ahead. Whereas in communicatory and perceptual abilities, such as combining television images with the narration, the bonobo was always advanced. The two species may have different cognitive strengths" (40). No one loses in this set of distinctions. There is no contest. There are simply differing strengths. And remember that chimps and bonobos represent different branches of that tree, branches that connect them to us more closely than to the other great apes—gorillas and orangutans.

PART II

Either men will learn to live like brothers, or they will die like beasts.

—Max Lerner, *The Unfinished Country*

I give you a message from this enigmatic, molecular world: . . . the cry of the beast:

. . . GGC ATC CTC AGC TAC AGG GTG GGC TTC TTC CTG . . .

—Simon Mawer, *Mendel's Dwarf*

In troubling the waters of kinship based on an economy of blood, it becomes apparent that other bodily fluids and substances remain rich metaphors and metonyms; they enable us to cross borders, as well as identities. Permeability raises these questions of intimacy, and what passes across membranes may result from the mixing or shedding of blood. While other mediums are available for the reckoning of kinship, blood still counts when we contemplate who deserves our care and whom we exclude from the regime of the familial, the familiar. These remain crucial sites of conflict and reconciliation in the making of social life.

Blood is what we spill, carry, and pass on to the future. Mapping some of the oldest and newest ways to reckon whom we care for and some of the ways that genealogies are scripted requires addressing our willingness to relinquish earlier modes of kinship so as to have a hand in shaping its very survival. In *Antigone's Claim*, where Judith Butler called to our attention the fact that kinship is a matter of "life and death," her focus was on the sister/daughter, Antigone; in this chapter, mine is on a particular set of agonistically entwined brothers.

Andrew Niccol's film *Gattaca* (1997) will serve here as an exemplary tale of intervention into a set of kinship networks that defy easy categorization. We can see in this narrative of an old story of "blood brothers" a scenario of biological kinship discarded and reinscribed in a "not-too-distant future." *Gattaca* is a tale of identity exchange and traffic written and directed by Andrew Niccol, who also wrote *The Truman Show* (dir. Peter Weir, 1998), *Simone* (which Niccol also directed, 2002), and, more recently, *The Terminal* (dir. Steven Spielberg, 2004). We now read the late 1990s as that moment when the questions of exploration and discovery were focused on the cataloging of the human genome. Although that realm was imagined to be vast, it finally measured up more humbly than had been expected in matters of sheer numbers. Nevertheless, countless new narratives have been spawned from this knowledge, producing novel tales of men and their chromosomal mice cousins. *Gattaca* is one such eu/dystopian tale.

In this film, blood relations are kept in play through a dialectic of the theological and the technological that is announced in two epigraphs: "Consider God's handiwork. Who can straighten what he hath made crooked?" (Ecclesiastes 7:13); and "I not only think we will tamper with Mother Nature. I think Mother wants us to" (from Willard Gaylin, "What's So Special about Being Human?"). This crookedness and tampering are telegraphed in one of the first scenes, where we see a refrigerator filled with meticulously arranged pouches of blood and urine. We learn that Gattaca is the workplace of astronauts. Following the credits, we watch a flow of arriving workers that restages Fritz Lang's *Metropolis*, with the futuristic fillip of machines that check fingerprints and draw fingertip blood samples as routinely as we now "swipe" bank cards. Beginning a seemingly ordinary workday, the director (Gore Vidal) of the Gattaca Corporation remarks on how clean our hero, Jerome (Ethan Hawke), keeps his workstation; in reply to his superior, he comments with understatement that may or may not be irony, "They say it's next to godliness." The term *Gattaca* is also a name extracted and distilled into a kind of acronym linking the four letters that represent the protein-forming bases that constitute the double helix arrangement of DNA: guanine, adenine, thymine, and cytosine. In the opening credits the letters *G*, *T*, *C*, and *A* appear in a bold font within the names of actors and others responsible for the making of the film. This not-too-distant future presents us with a tale of human efforts toward perfectibility, and the inevitable totalizing fears raised by early- and late-twentieth-century genetic narratives of dystopia. If Margaret Atwood's *The Handmaid's Tale* depicts a theocratic post-technological feminist dystopia, *Gattaca* stages its masculinist technophilic inversion.

Gattaca is an epic and elemental tale of Earth and the desire to leave it. In the voice-over of the early scenes, Jerome Morrow introduces himself by taking us back to his childhood dream of going into space. We are put on notice that there is deception in this regime of regulation of the body when the narrator tells us, however, that he is "not Jerome Morrow." In this cinematic tale of fratricidal love, our hero's name and identity are changeable, but his desire is bound from childhood to a dream of flight in the form of space travel. There is a long history of heroes who dream of flying: from Icarus to Milkman Dead, the hero of

Toni Morrison's *Song of Solomon*; Amelia Earhart to Mae Jemison, the first African American woman astronaut. Whether they succeed or fail, they often become subjects of myth. In this future, as in our present, being an astronaut is a prestige profession—witness the hagiographic impulses spawned by the loss of the *Columbia* shuttle in February 2003. Two months later, the first investigative reports—whose authors included Sally Ride, the first American woman in space, and Diane Vaughan, a sociologist and the author of *The* Challenger *Launch Decision*—addressed the problematic and, I would add, masculinist decision-making culture of NASA that demanded attention as much as "any falling foam or data recorder" (Schwartz and Wald, "Echoes of *Challenger*").

Consider this an exercise in interpreting some of the signs by which we can decipher the dialogue between past and future here in our present. As I remarked earlier, technology *and* kinship are what we make of that which we are given—the potential derived from the possible and available—whether fishing for termites like some of our primate cousins or going into space like those with the right stuff.

Both the opening echoes and the coda of *Gattaca* pay homage to Stanley Kubrick's landmark rewriting of the past into the future in *2001: A Space Odyssey* (1968). Recall the thigh bone in slow motion hurled upward and transformed in silent descent into a spaceship set to the music of Richard Strauss's *Also Sprach Zarathustra*. Niccol opens with Michael Nyman's original score accompanying falling objects that look very like that femur but turn out to be magnified nails, hair, and skin, as we learn when the camera pulls back and we see those body parts being shed in a compulsory, disciplined manner by an anonymous male bather. We glimpse him from a variety of angles as he emerges from his shower, and, with the flip of a switch, water turns into fire as his sloughed-off body parts are incinerated. This is our narrator, our "navigator, first class," getting ready for a day's work. We might wonder, following anthropologist Mary Douglas, about the pollution suggested by things-out-of-place, a pollution that is disclosed when he opens that refrigerator door and removes a urine sample and straps it to his thigh. This is obviously a system of intensely monitored bodily hygiene, but it is not one we recognize. The tension between the familiar and the unfamiliar circumscribes the temporality of this not-so-distant future. Janet Carsten has described kinship as the relations that cultures establish between substances and the sentiments they engender. The carefully stored pouches we saw contain the precious fluids of a market where blood is exchanged and purchased—where vital *substances* have been radically severed from *sentiment* rather than functioning to signify their alliance. In this future, blood and other fluids constitute the materials of an underground economy of identity trade. Substances are renegotiated and transvalued in a system that circumvents the risky business of a continuing genetic wheel of fortune where some who cannot win purchase their way to prestige.

The long and sedately paced introduction to the film is followed by an even longer expository flashback to a young, white, married couple who have decided against conceiving their second child as they did their first, the old-fashioned way (in the backseat of a Buick Riviera), and instead seek out genetic counseling. "You'll want to give your child the best possible start," is the sensible advice of the brown man in medical white coat (Blair Underwood), who helps them scrutinize their list of desired traits. To quell their lingering doubts in offering a genetic makeover, he points out that "you might conceive naturally a thousand times and never achieve such results." Now they must choose one among four viable embryos. And because there are two males and two females to choose from, it appears the only real choice is the sex of a sibling. "Of course, we would want Vincent/(later Jerome) to have a brother," says his mother (Jayne Brook), with a minor hesitation, yet making her family into the image of her heart's desire—investing substance (multiplying cells) with sentiment (the love of brothers), mixing these ingredients in kinship's crucible. The wise counselor adds, "He's still part of you both, just the best parts."

We see the brothers, Vincent and Anton, as boys, their sibling rivalry tested by who can swim farther out into the ocean. Vincent is the elder yet weaker brother; he is what is called a "faith" or "god child," conceived in love, "once thought to be the basis of happiness." Anton (Loren Dean), the younger, is the "son worthy of his father's name." He is the one conceived with the assistance of the local geneticist, and giving this child the best chance has come to mean *not* leaving things to chance. One day Vincent is finally and surprisingly the victor in their competition. Vincent's 99 percent chance of heart disease, accurately predicted at birth, and his life expectancy of 30.2 years lead his father (Elias Koteas) to say to his adolescent, space-obsessed son, "The only way you'll see the inside of Gattaca is if you're cleaning it." What hero could have asked for a more provocative paternal curse? Having successfully defied the cardiovascular odds, his turn-of-the-twenty-first-century picaresque journey begins. He becomes part of the new underclass, those less genetically sound. When we next see Vincent, he is, in fact, one of the men in the custodial crew at Gattaca, assiduously studying its details of access and hierarchy. The earlier vanquishing of his brother has left Vincent certain as never before that he can and will continue to pursue the impossible.

The day for which we have been watching him prepare and dress for work is the day his lifelong wish is to be fulfilled. Director Josef has come to his workstation not to make small talk about cleanliness and godliness but to announce that Vincent has been chosen to leave in a week for Titan, a launch that can be attempted only every seventy years. This crack in the window of cosmic opportunity fuels a subplot because on that same day the mission director (a character we see only as a corpse) is murdered. The investigation that follows threatens to reveal that Vincent is passing as Jerome Morrow, a "valid" citizen of this future. The lead detective on the case is that brother left behind long ago, Anton, a young professional who insists on the most up-to-date methods, accompanied by his more senior yet assistant colleague in an old-fashioned fedora (Alan Arkin), who at a certain moment suggests "blood from the vein" when urine, saliva, and fingertip specimens fail to uncover the identity of the murderer. With Vincent/Jerome's new/surrogate father determined to launch the mission to crown his career, and the former/birth brother as the young lead detective on the murder case, we witness the intertwined struggle between brothers that, as always, invokes the fathers in this rewriting of a new totem and taboo.

The fraternal agony of the family of origin was literally underscored at the moment when Vincent tore his face from a family photo and left home, never to return. To become an astronaut-in-training, his identity was disappeared into the genetically superior but physically disabled body of Jerome Morrow. Where Vincent and his blood brother, Anton, had tangled in the ocean with their regular swimming contests, Vincent and Jerome share bodily fluids to facilitate the desired trajectory into space. They meet through an intermediary who vets Vincent's commitment to this genetic slipping of the punch. In this auspicious meeting, Vincent is admonished by his genetic broker, German (Tony Shalhoub), that "there is no gene for fate." When we finally meet the real Jerome Morrow (Jude Law), we see that he is confined to a wheelchair. Although he holds a silver medal in swimming, this perfect human specimen, who was a winner in the genetic sweepstakes, is bitter because he "was never meant to be one step down." We learn later that he lost the use of his legs after he stepped in front of a moving car—a failed attempt at suicide because he had not brought home the gold. The most cynical character in this film, he will, in spite of himself, find a new purpose in letting Vincent take over his identity inasmuch as his fluids make it possible for Vincent's body to leave Earth while he cannot leave home unassisted. We come to read this pair of Jeromes as an amalgam of a queer coupling and a chosen twinning of the genomic age.

In *Gattaca*, God and the local genetic counselor are on speaking terms. If the questions of Kubrick's cinematic epic of 1968 addressed the struggles among those men and their machines sent into space to do the work of dividing and conquering, the questions of 1997 circle back to who, among those on Earth, gets to go out and up there. To pass as an astronaut, an occupation of the genetically entitled, Vincent must become, in the idiom of this time and place, a "borrowed ladder." He will impersonate Jerome Morrow in order to vault over the deficiencies determined at his birth. Vincent is a member of a vanishing species, so to speak, those whose births were left to chance rather than to the efforts of good breeding—that is, eugenics.

While we learn that discrimination on the basis of genes is illegal in this future (it is called "genoism"—the first two syllables rhyme with the city in Italy), we also know there is a black market in precious substances because "they've got discrimination down to a science" and your "resume is in your cells." Identities are nonetheless exchanged in a system where identification is routinely verified by regular urine testing, saliva collection, and the "hoovering," or vacuuming, of workers' keyboards and desk drawers for stray hairs or flakes of skin. Later a single eyelash will threaten to unmask the perpetrator of the murder that serves as the plot engine.

Gattaca, a thirty-year aftershock with its echoes of *2001: A Space Odyssey*, is yet its own story of struggle between brothers—a struggle as biblical as it is postmodern. Here blood is not shed but traded. The film proffers an ameliorated present, one inevitably fraught with the undercurrents of the past. It is declared that "blood has no nationality," and the workforce at Gattaca incarnates multiculturalism made fact. We do see men and women of color in positions at the higher ranks, yet those in command remain able-bodied white men. Vincent, the imperfect son, aims his life toward becoming the chosen son—the one to prove his biogenetic father wrong. Instead, he will do his best to consummate the visions of two of his father-surrogates: Director Josef, who has graced him with the mission to Titan, and Doctor Lamar (Xander Berkeley), who performs the regular workplace substance tests and repeatedly alludes to a son we never meet and who has known and kept Vincent/Jerome's identity secret. In a very early scene, when Jerome gives a routine urine sample, Lamar remarks, "Never shy, Jerome. Always pisses on command. Have I ever told you what a beautiful piece of equipment you've got there. I don't know why my parents didn't order one up like that for me." The struggle with what is possible or impossible sets the entire movie in motion. We are told that no one "exceeds their potential" in the future of *Gattaca*'s making, and if they do, then there must have been a miscalculation of their potential. Or collusion.

The technology of care of the self reveals a biopolitics where the body's most ephemeral parts (nails, hair, skin) are recalculated as the repositories of the most classified information. When we first saw Vincent/Jerome, he was performing his daily bodily cleansing routines, which included disposal of these quickly renewable parts in an autoclave—a device of such heat as to consume the traces of Vincent so that he may live as Jerome. The medical facts of Vincent's life are subject to manipulation and falsification. With the daily disposal of his bodily detritus, and re-supplementation of the proper blood, urine, hair, and skin samples, Vincent will live with the new brother he has chosen to impersonate, and together they will obsessively master this regime of sharing bodily substances.

Vincent, having taken Jerome's name and place, becomes a "valid" citizen rather than an "invalid" (accent on the second syllable) in the language of this universe. Like all eu/outopias, language tells the tale economically: he has gone from being a "degenerate" (the second syllable pronounced "gene") to someone who can now go "anywhere with this guy's helix wrapped under his arm," as he is told by German, the identity trader. A new self is born who can pursue the dreams and desires of the old self, if provided with daily blood, urine, fingertip samples, and hair samples so as to keep his former self obscured. A new brotherly love develops between the two Jeromes as they engage in the process of shedding and mixing selves. The British-accented and wheelchair-bound Jerome suggests that Vincent now call him by his middle name, Eugene (a name popularized in the late nineteenth century and early twentieth, after Eugene Galton, cousin to Charles Darwin, directed the quest to understand evolution in the popular direction of eugenics). In an apotheosis of instrumental rationality and unsentimental needs, bodily substances will ultimately return as sentiment when Vincent and Jerome have succeeded in becoming two-as-one. Like lovers in earlier centuries, the bond between these brothers is memorialized when the gift of a lock of Jerome's hair is what Vincent takes into space on his mission to Titan. Should the weak-hearted Vincent survive the launch, he will at least come supplied with a repository of the DNA that might entitle him to continue to pass as a proper member of this mission to the stars.

The queering of this couple, these brothers under the skin, is underscored by the fact that women are notably scarce in this story: there is Vincent's naive mother, who insisted, at his birth, in opposition to all the negative medical indicators, that "he'll do something." And, years later, there is Irene (Uma Thurman), a co-worker who becomes suspicious of Vincent/Jerome's avid preoccupation about watching every liftoff. Irene is the barely necessary heterosexual dressing for what remains a scenario of masculine struggle and striving. (Earlier films of space travel, particularly *The Right Stuff* and *Apollo 13*, while completely focused on the desires of men, nevertheless went out of their way to portray, if only briefly, the wives and women who struggle to be partners to such desires.) To find out more about "Jerome," Irene filches a hair from the comb in his desk drawer and takes it to the genetic bank, where we see women stand in line, offering up bodily residues and fluids to be tested, and leaving with printouts of genetic profiles. Later she confesses to him, "I've had you sequenced."

It will turn out that his secret self—the one wearing contact lenses, strap-on urine samples, and blood sachets on his fingertips—is safe with Irene, for she, too, has made it to Gattaca by dint of hard work and has exceeded her origins, which are, like his, those of a "god child" with a weak heart. Their dilute romance is forged out of their shared history of having both been born out of an earlier equation than the one that powers this imaginary present. With just a few days left before being launched to Titan, Vincent/Jerome shifts into a party mood. First he and Eugene/Jerome go out on the town, an evening that culminates with Vincent's carrying his drunk and disabled (br)other to bed. Another evening, while out with Irene, the two of them are stopped by police seeking potential suspects in the murder investigation; the traffic stop includes cops behaving like doctors, asking drivers and passengers to open their mouths as they swab for saliva samples. Vincent/Jerome quickly tosses out the car window his contact lenses that would be revealed by flashlights and could lead to his being discovered to be an impostor, an invalid.

The tightening of the investigative net leads to a scene of classic Hollywood (mis)recognition among four of the players, first an encounter between the hero's biological brother and his brother of the good genes. Vincent is alerted by Irene to the fact that the detective/brother is heading to his home; he, in turn, phones Eugene/Jerome to tell him he must "impersonate" himself for this impromptu interrogation. Irene and Vincent rush to the scene. The authentic/real Jerome Morrow, alone at home, drags his paraplegic body up the spiral staircase (a double helix structure), properly sits himself in a chair, brushes dust from his pants and sweat from his brow, and answers the detective's questions with aplomb. Irene arrives just after the detective, and staring at a man she's never seen before, who plays the part of her suitor asking for a kiss, she gives him one. The detective leaves, and the man with whom she has just spent the night arrives and begins to explain. Irene thought she knew all his secrets. But with this last imposture exposed, she storms out, and Vincent is downcast. Irene leaves, dejected by the ruse she had failed to fully understand. Eugene tries to cheer Vincent up, saying, "I think she likes us." A brotherly love could not speak more clearly; their twinning is complete in Jude Law's ironic delivery of the line.

The focus on the brotherly discloses an economy of masculinity that holds sway in this future. Vincent has chosen new "fathers," as well as this new brother. First, there was the boss on the cleaning crew (Ernest Borgnine), who tried to cut Vincent down to size when he saw his fascination with the astronauts and technocrats. There is, of course, the director, who has chosen him for the mission; it emerges that the director is in fact the murderer who insisted that this mission go ahead as planned. With a launch possible only every seven decades, he will not live to see another. His own dream must come true, and Jerome must live it. And finally, there is the doctor, Lamar, who has known and kept Jerome's secret identity. This revelation occurs in Vincent's last unscheduled urine test before liftoff, when Lamar casually points out that "right-handed men don't hold it with their left." This is the moment when Jerome recognizes that he has been helped in his path toward fulfillment not only by the bodily substances of his "double" but also by the desires of others who vicariously feed off his dream. He is their hero, as well as the hero of his own lonely quest.

The melancholic musical echo that opens the film is the sound of the past resonating through the future, into it, and interpenetrating it. The image accompanying that echo is defamiliarized in the beginning—those beautifully raining flakes of skin, strands of hair. But the music returns over the closing credits, where we see anew the magnification of nail clippings, shaved hair, and scrubbed skin, the last vestiges of the self being cast off, shed by the body that married Vincent's flesh-and-bone form with Jerome's cellular structures. Where the protohominid of

Kubrick's 1960s vision hurled a bone up into the air, a bone that silently transformed into a spaceship, Niccol lets the pieces fall back down, reversing the image and turning up the volume, posing questions about where and how far we have traveled since. Leaving our watery and earthly realm, both Jeromes are propelled by fire—one brother is rocketed into space, the other cremated in the sterilizing blaze of hearth and home they shared as one mixed identity. A final ritual of transubstantiation merges the techno- and theological scenarios.

Vincent was an outcast in a setting where neither the mother's body nor the father's name guaranteed who would be his brother. When they meet in the course of the investigation, we see the brother his father elected, Anton, and his mother imagined as "someone for Vincent to play with" in a recognition scene where the elder asks of the younger sibling, "*Are* we brothers?" The only play that was ever possible between them was the rivalrous bodily testing kind. Since they share the same blood, we imagine they might not ever have bothered to engage in the childhood ritual of pricking their fingers and mixing their shared substance. But Vincent's voice-over explains his own long-standing doubts that they were made of the same "stuff." Anton had the right stuff, but it was Vincent who dreamed of space travel. As adults they replay their swimming challenge, and Anton gets to ask how Vincent ever managed to win. Vincent explains that he never planned for the return trip. When he erased his name and height from the family score-keeping game, he disappeared into the new underclass of the genetically impoverished. Anton clarifies that their parents died believing they had outlived Vincent. In those intervening years, he had continued his autodidactic ways, mastering the details of out-of-date astronomy textbooks and stolen moments at the computer screens of Gattaca's privileged workers, while Jerome Morrow, a winner in the genetic sweepstakes, was accidentally disabled.

For them to become brothers, they had to renegotiate their abilities and disabilities. In a key scene, with Jerome in his wheelchair and Vincent on crutches—recovering from surgery to make him as tall as Jerome—and neither one standing on his own two feet, Vincent notes the irony of the fact that in space, "where I'm going, you don't need legs." When Vincent thinks he'll be caught passing, Jerome/Eugene, his more perfect disabled self, reminds him of the regime in place: "They don't see you. They see me." Fluids reveal truths, not the eyes. Among the brotherly rituals established by these two is their careful attention to the blood and urine stored for daily use. An "odd couple" moment has Vincent nearly late for work because Eugene has been drinking and all the urine specimens are "hot." Although bodily substances are repeatedly tested for authenticity of identity, even under such ubiquitous eyes piracy is practiced.

Both Jeromes leave the earth. Eugene's death wish, frustrated by his accident, let him live on Vincent's dream. The Olympic athlete who learns literally to dance with his wheelchair was spared the successful suicide attempt and instead shared a bodily life with the weak-hearted boy who anticipated flying. Some find the mood and meaning of *Gattaca* macabre. While the film maintains a rhetorical tone and a mise-en-scène that is consistently chilly, it also hints at an antic side, particularly in its casting and understated humor. Niccol's vision is neither an unqualified utopia nor dystopia; rather he asks viewers to enter a scenario that poses all sorts of questions about identity and desire and offers only some speculative answers to where curiosity and knowledge could take those of us on Earth. I find in Niccol's film a hope that stages what might be possible if bodily boundaries were even more permeable than they conventionally are among those we think of as couples, siblings, parents, lovers.

What was a "not-so-distant future" in Niccol's 1997 is, at the time of this writing, in 2007, a time when the information available through genetic testing pre- and postnatally expands daily, and privately funded "manned" space flight has had its first liftoff. And from the *Cassini* "unmanned" spacecraft, images return to us on Earth from the moons of Saturn, including "close encounters with Titan." What constitutes the human of the twenty-first century is certainly the desire to straighten what is crooked, and tampering with nature comes naturally to us. While the intimacy practiced by these two men is initially contractual and conflictual, it is finally a "kintimacy" where each has dreamed the other's dream and saved the other's life.

The evidence of how deeply kinship pervades bodily states interrogates the phenomenological, the sensory, and questions of boundaries. We know the idioms of things getting under our skin, our truisms about blood being thicker than water. Kinship is one of our "cultural naturals," Frans de Waal reminds us throughout *The Ape and the Sushi Master*. My reiteration of the variability and mutability of what we make of what is given to us points toward the permeability of realms of intimacy, forms of kinship. And we know that these are ferociously contested sites where we cling to those kinship narratives we proclaim.

By foregrounding the metaphor of "kissing cousins," I want to insist on a malleable arrangement, one that might suggest models of kinship for this century, models where permeability creates a demilitarized zone of kinship that could displace the familiar logics of blood. Consider how widely available the term *cousin* is for naming proximity and relatedness—intimacy and distance. It is common-place to allow it to cross even species boundaries. Ever refined "out of Africa" theories of origins repeatedly speak of our closest primate relations as cousins.

Kinship becomes a language, a cultural idiolect, a set of practices with a grammar, a rhetoric, even a poetics, as Margaret Trawick suggests. And Evelyn Blackwood points out that the "new kinship returns as an idiom of social practice, a set of ideas—about relatedness, 'blood,' sociality, friendship, obligation, and origin—that give meaning to social relations" (*Webs of Power*, 16). Our fluency in the language of kinship depends on our abilities to translate across the membranes of systems of classification. And de Waal reminds us that "we usually don't exclude the taming of fire from the cultural domain simply because it has been achieved by every human society. Some cultural inventions come naturally to us, such as building roofs over our heads, performing marriage rituals, or developing a classification system for close kin" (*The Ape and the Sushi Master*, 285). Whether those roofs resemble tents, caves, or nests, there are clear indicators to those in the network about who belongs and who must be forbidden entry.

If kinship is indeed a matter of life and death, what is revealed when we read the world we live in as driven by the interlacing, interlocking struggle over blood? At one time the gold standard became an epistemological marker in the understanding of economy. For an overly long time the currency of blood has operated as the liquid asset of histories of conflict and reconciliation. Those old codes of blood are displaced and replaced in the twenty-first century by newly legible bodily fluids and solids. The policing, storing, matching, and trading in such substances prescribes anew our relations with those we consider intimates and those who remain outsiders. If the question of whom we may kiss were answered more forthrightly than whom we may kill, the latter category could shrink with the expansion of the former. Just think of the many places where we watch the minimal differences of kin explode into mass scenarios of death, not birth; of murder, not marriage; of barriers going up, not coming down; of blood shed, not mixed. If we could think and live this kissing–killing knot with empathy as the starting point, we might be able to move forward into our mixed and mingled futures with forms of radical hope fashioned out of old histories of despair.

"Do we have to talk about this?" says the exasperated and about-to-be-outed nonbiological father of Leon in the British film *Leon, the Pig Farmer* (1992). The film stages a scenario of the unintended consequences of anonymous donor insemination in a closeted world of heterosexual infertile couples. A recently established mother-and-son-designed Web site, The Donor Sibling Registry (donorsiblingregistry.com), fuels the same desire that animates Leon, a North London, middle-class Jewish boy, to go on a quest for his biological father when he discovers through an "accident"—a term for unplanned children—that his origin lies somewhere other than in the nuclear family within which he has grown up.

Leon's origins are shifted and doubled in the story that follows him back to his biological rather than social origins—the home of a Yorkshire pig farmer. Having been raised by kosher caterers, Leon is haunted by the porcine paraphernalia of the rural household. This married couple eagerly try to make him more comfortable by studying up on the habits and customs of Jews, who are an alien species out here in the countryside. Leon's donor-father attempts to integrate him into this family business, teaching him some of the methods for breeding the best pigs. When the local veterinarian makes a "mistake," producing an accidental beast half-pig, half-sheep, Leon's own conflict is replayed in the animal realm as he intervenes in the survival of this half-breed.

The entire film is preoccupied with questions of tribal loyalties. Will he marry a nice Jewish girl, and soon, please, as his family nudges him both in his imagination and at events where extended family members get to chime in on his single state? Why not consider, says one of his friends, that "opposites attract"? "What are we, magnets?" asks Leon of his friend. Before too long a woman he meets when his car hits her bicycle is having him pose for her latest stained-glass artwork—a crucifixion scene in which her exotic Jewish lover plays Christ.

Do we have to talk about this?

I'm afraid we do. Because reproductive technologies that are far from "new" remain one of the earliest and most abiding sources for the new forms of relatedness and kinship being produced in this postmillennial moment. Whether the theories of some evolutionary biologists are correct in proposing that women at their fertile times are most likely to stray into affairs, thereby choosing superior donor-fathers for their desired children, or whether women are agents of choice as customers of sperm banks for various reasons, it is clear that women and men certainly intervene in the methods of reproducing those of our kind. And our interventions are becoming more varied in their approaches to the desire for children.

The boundary waters where we acknowledge and simultaneously deny our kinship with our animal cousins—

these are where we come to wade here. The primitive and the futuristic keep meeting up. Farmers like Leon's biological father breed their pigs for the most richly endowed characteristics that lend them value; yet the notion that Leon's mother and social father might have also availed themselves of proper seed for propagating their family comes as a shock to Leon and makes/breaks a taboo of silence with his social father when Leon reports the news of his discovery.

Leon, the Pig Farmer takes a close look at tribal affiliations and their insistence in lives chosen, not fated. Leon must come to terms with his encounter with the clichés about whether opposites attract or whether one must stick to one's own kind. There is a mixing of metaphors and a mixing of kinship lines that is played out in his rescue of the hybrid pig–sheep. He rescues it, only to free it in the woods somewhere between Yorkshire and North London. But this *mischling*, this chimera, won't find another of its own kind in those woods. No, it will forage and sometime in its undoubtedly short future will come to grief alone or be savaged by beasts that don't recognize it except as prey. Leon is its hero, its righteous human who doesn't want to see it euthanized for its hybrid nature but who also doesn't know how or where it should survive. Leon feels a kinship with this unnameable creature that mirrors his own confused search in the wilds of Yorkshire for how to return to his family of origin in London. This search is one that is familiar no matter the genealogical and genetic individual histories being constructed through Web sites or fertility clinics.

This story exemplifies in its particulars some of the comic and poignant narratives being lived in our time. In *The*

Genius Factory, David Plotz comes at these stories from the perspective of a journalist tracking down anyone who may have been involved in making use of the so-called Nobel sperm bank, which functioned from 1980 to 1999 and produced some two hundred offspring of anonymous donors who were reputed to be "geniuses." This was an institution founded by men who had visions of a genius race to outwit and one day outnumber the overpopulation of nitwits they saw around them; and it ran its business in tandem with its customers, women who made use of the sperm available to make families. Plotz tells the story of one of the first times that "sperm bank half siblings" meet; he notes that although they have DNA in common, "their shared father is a complete blank" (63). The meeting of two young men—"sperm bank brothers"—offers a moment in which the "only thing they knew about him was that they didn't know anything about him." This anonymous donor-father, known as Coral, the kind of name used by the "genius factory" workers, represents what Plotz calls a "paternal void." One of these "voids" recently made himself known; he was called Donor 150 in an article that featured the results of the Donor Sibling Registry when several offspring of the same anonymous donor met to satisfy their urge to know something about their paternal void—but what they came to know was each other and their families of mothers and siblings. But more recently, Donor 150 filled that void with his own history of having been a regular and much desired donor at a California sperm bank more than a decade ago. As "anonymity crumbles" in the days of the Internet, Plotz notes that "in sperm shopping, there is a deposit, but there are no returns, no refunds, no exchanges" (181). Sperm banks may have seemed revolutionary a few decades ago. But in our time, the interventions into known and unknown forms of kinship have taken the issues and questions far afield through the work of science and the imagination.

I'd like to move from these social tales of reproductive recipes for the making of families to the arena of art, where science and desire take rather stranger forms than the very familiar wish to know one's origins or reshape and share one's origins. Patricia Piccinini's work was highlighted at the Australian pavilion of the 2003 Venice Biennale, but I came to it at the group exhibit "Becoming Animal," which ran during 2005 and 2006 at the Massachusetts Museum of Contemporary Art. With her recent work in the realm of the visual plastic arts, Piccinini poses questions similar to those that have concerned me here, especially with her work *The Young Family*. This family is described as a hybrid animal–human mother suckling her young—they appear to be human and piglet-like in their bodily attributes. What is represented, however, is the fact of her birthing and rearing young who are clearly her own offspring—we wonder which opposites attracted each other in the lab where she may have been produced and are now reproduced in a younger generation. There can be no doubt of her maternal devotion and her weariness in her immobility—her young at her teats and belly. The catalog editor writes, "While her babies feverishly suckle, her tired and weathered eyes provide an all-too-human window to her soul" (Thompson, *Becoming Animal*, 98).

In an interview with Nato Thompson, one of the curators of the MASS MoCA show, Piccinini talks about the gaze

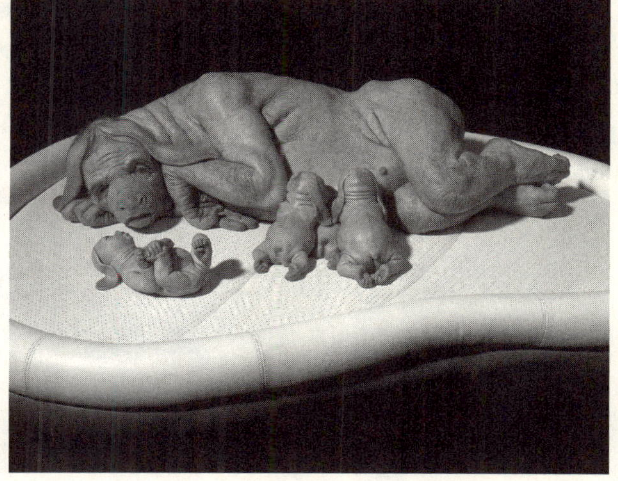

The Young Family. (Photo courtesy of Patricia Piccinini)

of her hybrid subjects, about flesh becoming plastic, and about the irony of working in silicone: "a kind of plastic—to create flesh in works that talk of the plasticity of flesh" (quoted in Thompson, *Becoming Animal*, 104). She also acknowledges the introspective nature of her subject's gaze and says that animals "are less interested in us than we are in them" (104). As I have been aiming to do in these chapters, Piccinini further notes that she uses animals "to tell stories about the world we live in or to try to explain or explore ethical issues that are important to our times" (104). The border between selves, human and nonhuman, is a site of exploration, as we've seen in multiple media and in the stories of everyday lives. Encountering Piccini-ni's bestiary allows us to confront some of these borders in the realm of art, where the artist may be able to pose the ethical questions more visually than rhetorically, imposing on spectators the time for reflection on the interventions and discoveries from the realm of science that hurtle us forward as our visceral responses and intellectual debates attempt to keep pace.

13 MENDEL'S NEPHEW

If truth is indeed stranger than fiction, consider some of the kinds of truths we must imagine going on around us—in vitro racially mixed twins-separated-at-birth as a postmodern rewriting of the potentially tragic consequences of mixing. Simon Mawer, a British novelist, calls such stories contemporary urban legends, but court records involving adoption of mixed-up embryos attest to the real in the legendary. Here, for our perusal, is a recent fiction by Mawer that enthralls and appalls as it puts before us the element of choice in the house of fate, to paraphrase Jennifer Ackerman.

Ackerman's suggestively titled book *Chance in the House of Fate* asks us to think about what becomes (chance) of what is given (fate), another spin on the "cultural natural," an area of reconciliation and conflict. I would like to consider such negotiations as subsumed under a notion of kinship where cousins may or may not "kiss" and, in the decision made, construct ideas of permeability. Working through our desired and dreaded new narratives of reproduction, Ackerman reminds us that "identical genes do not make identical people" (44). She demands that while we may exercise ourselves over the depredations of current and future scientists, we ought not to forget the "monstrous or miraculous" (44) deliveries of old-fashioned techniques of reproducing ourselves.

The protagonist of Simon Mawer's novel *Mendel's Dwarf* is Dr. Benedict Lambert, an achondroplastic (dwarf) geneticist and the great-great-great nephew of Gregor Mendel, the Moravian botanist who codified the mathematics of genetics. The novel opens with Dr. Lambert as the plenary speaker at a conference of geneticists in the Czech Republic town of Brno, the town of Mendel's birth. The conference has assembled to discuss the latest twentieth-century genetic research. All eyes are on *him*, the speaker, but the reader sees through his narrator's eyes the scientists assembled in his audience who share "a constancy that is obvious to all, but consciously perceived only by the truncated figure up on the podium: each and every one of the earnest watchers is subsumed under the epithet *phenotypically normal*" (2).

Mendelian genetics, the kind that comes to matter more and more now that we are reading the map of DNA, is the science of how invisible parts make new wholes. Jean, the librarian at the research institute where she and Ben work, has an affair with him and, what's more, conceives. The potential genetic outcome of their segregated but penetrating codes becomes Ben's obsession. His work becomes the echo of his great-great-great uncle's assiduously kept register of the mixing of peas and fuchsias. The idea of becoming a father—so previously unimaginable to Ben, and now perhaps inexorable—leads readers through a tour of possible recombinations of what Ben would want for his imagined offspring. What does he desire of the chance thrown by fate that might be born of his love for Jean? How will his knowledge and power color his fateful interventions? Scandal is certain—whether from Jean's estranged husband, who is given to violence, or from Ben's critical professional colleagues.

Mawer is fluent in the discourse of science and lets that discourse fuel a narrative of conventional romantic folly between our scholar, Dr. Lambert—Ben—and a librarian, Jean Piercey (even as narrator, he apologizes to readers for her overdetermined name: "Don't laugh," he pleads [72]). Jean and Ben knew each other before his career brought him international attention. Ben's experiences with women have occurred almost exclusively through representation: magazines, videos, the occasional exchange with a woman who works at sex, and, more than anything else, his imagination. But Jean was one of two young women who were the objects of Ben's adolescent lust; her perch on the librarian's high stool put his eyes level with the fascination of the darkness beyond the triangle of her skirt just barely exposing her knees. Ben is an aficionado of knees, he tells us, since those are the body parts most visible to his eyes.

When Jean comes to Ben for sympathy about her infertile and abusive husband, they end up making love, and from this encounter comes a pregnancy that she decides to abort. When she asks naively what they will do if she gets pregnant, Ben enlightens her thus: "the chance of *me* happening was one in fifteen thousand. And here I am. Chances are things that have a habit of happening" (168). Only later will she come to him for help in keeping her marriage alive by the conventional choice to bring a child into her life, asking him to be the genetic father. After all, by this time he has discovered the gene for achondroplasia, so she assumes he will be able to determine whether her child will be "normal": "Benedict Lambert is sitting in his laboratory playing God. He has eight embryos in eight little tubes. Four of the embryos are proto-Benedicts, proto-dwarfs; the other four are, for want of a better word, normal. How should he choose? . . . But Benedict Lambert has the possibility of beating God's proxy and overturning the tables of chance. He can choose" (238). However, we don't learn by the end of this chapter, "Antibody," what he has chosen.

If kinship names what we choose in relation to what is given, then Ben's ethical dilemma is what he will make of what he thinks he knows. What does he desire of the chance thrown by fate that might be born of his love for Jean? The odds are fifty–fifty that his child will be, as he regretfully agrees to the term, either normal or a dwarf. Mawer teaches us that though there is no gene for grandmother's nose, per se, a mere fold in a DNA chain will make Ben's offspring look just like his father in one particular/particulate way. Jean and Ben have played games with their fate.

In the realm of reproduction, we have seen how what distinguishes us is our capacity to bring choice to chance. Recall the act of the narrator of Jeffrey Eugenides's short story "Baster," who intervened to subvert the potion of fate, chance, and choice. When his former girlfriend gave a party where she got to choose from among the men present the donor whose sperm she would use to inseminate herself later that auspiciously chosen evening, our narrator switched sperm samples in the bathroom.

What Mawer and Eugenides put before us as readers are scenarios of sexual selection that veer far off course from the more glacial evolutionary pace. In each of these fictions, men meddle in the course of a life history of a woman who has also daringly intervened in her own reproductive strategies in ways that would have been unimaginable before the closing decades of the twentieth century.

Mawer's narrator, who has acknowledged that the eugenics of the early twentieth century had led "to the ovens at Auschwitz" (224), now warns his audience of experts that the "new eugenics, *our* eugenics, is governed only by the laws of the marketplace" (273). Benedict Lambert insists on the word, as Mawer insists on reminding his readers where knowledge is led by ideological demands. And this time around, it is "all masquerading as freedom" (273). Ben is giving his lecture in the Czech Republic as Jean is delivering their baby in England; her husband still believes her pregnancy was the result of in vitro fertilization by his own (radically deficient) sperm. When he learns he has been duped through an act of "cytoplasmic cuckoldry," the consequences are tragic. With all the most current knowledge in Ben's mind and the desire for paternity in his body, Jean nevertheless succumbs to the very old and still possible fate of death in childbirth. In his grief, Jean's infertile husband, Hugo, wants to be sure this child is his. He has learned his Mendel enough to know his child ought not to have been born with brown eyes. He suspects the infertility doctor, certainly not dwarfish Ben, of having betrayed him. And Ben takes his revenge in words: "*I am the father. Ridiculous Ben Lambert is the father. Adam is ours—Jean's and mine. Nothing to do with you at all.*

Do you understand that? He is nothing to do with you. I was her lover and you were too damned prejudiced to realize it" (284). After Hugo leaves, Ben confesses to a "feeling of mild elation . . . the plain feeling that I had *won*" (285). But Hugo had told Ben that if the child was not "his," he didn't want it. And rage, the rage of betrayal—social, sexual, genetic, generational—leads to Mawer's dramatic conclusion. "From Mendel to the future: the tenuous chain of descent, the passage of DNA down the generations was soon broken" (293).

This is not simply a novel of romantic chance and fate, however. It is a novel accompanied by genetic diagrams, by footnotes to scientific debates, articles, and Web sites; Mawer leads his readers through a study of genetic knowledge in its century-long arc, one that ends before the millennial decoding of the genome. There was in 1998 still the presumption of a kind of complexity to the human genome that would turn out more modestly and humbly when 100,000 was revealed to be closer to some 30,000; still, the complex sciences about which we all now receive daily news need to be taught.

Mawer moves between Mendel and Darwin and the differing natures of their research. Mendel's years of data gathered through his "children," the peas and fuchsias and other less successful species, showed that "his simple mathematical model held true: inheritance was governed by particles, one contributed by each parent, no mingling of blood . . . one bead from each parent for each inherited character" (221). Darwin recommended Mendel's work to the author of an encyclopedia article on his own work. Ben, the narrator, insists, "Oh yes, indeed, Darwin *needed*

Mendel. . . . That is, he held that offspring tend to be a *blend* of their parents' characteristics. The trouble is, he was wrong" (228). In an effort to explain the basics of his profession as a geneticist to laypeople, and to defy their expectations of him or their efforts at politeness or political correctness, Ben is given to saying "I collect dwarfs" (116).

Benedict's ancestors include not only Mendel but, more recently, one Gottlieb Weiss, who fled Austria, came to England, and with his entrepreneurial spirit created a freak show. In going through the personal effects of a newly deceased uncle, Benedict comes upon the advertisement for "Gottlieb Weiss's Anatomical Curiosities" (57). This was the family venture until 1914. By the time the Great War was over, Weiss had metamorphosed into Doctor Godley Wise, Confidant of the Crowned Heads of Europe, and now a lecturer on the "Science of Human Genetics, founded on the new Mendelian Principles, being a Full Exposition of the Danger faced by the British Race through a Deterioration of its Genetic Stock" (59). This thread through European history as it elaborates Ben's family history, and his unique place as the most current authority in his field and in his very being, gives Mawer the opportunity to enrich his narrative with humor, irony, and the questions of utmost importance for the newest turn taken by knowledge that had previously been used to entertain and to annihilate.

14 OF LOVE AND LAW

This is a book about who counts. And who cares. Who is cared for by whom? Who are the members of the group that count for each other? And what are their caring obligations? In first being cared for by our kin, we become fluent in the ways of caring. Fluency is always acquired through trial and error. Who decides, for example, when the infant we hold must be let go so that she or he may learn to walk? And what is the affective and cognitive medium of this apprenticeship?

Recent writings in conflict resolution lead us to contemplate some of the links between law and love. Some have suggested that the prototypical scenario in which we play out conflict and reconciliation is the case of weaning. Surely this is a site where kinship relations and their obligations are unmistakable. Mothers must calibrate and ultimately titrate the bond necessitated by a prolonged infancy. In these ongoing deliberations of everyday life, relations of intimacy grow—relations involving the inevitable conflict that the laws of change enforce and the love of those nearest resolve in the courts of kinship. In "Law, Love, and Reconciliation," Douglas H. Yarn notes that "law is not designed to end conflict and is best categorized as conflict management rather than conflict resolution" (57). In early European courts "lovedays (*jours d'amour*) were an established institution for reconciliation" (60).

We humans do take our arbitrations of kinship to the law—and we anguish in hope and fear that the law will produce the justice of love. The mother who is weaning her infant must relinquish her love to the laws of separation and autonomy. Her law and her love are constituted in those very laws, and in them rests her and her child's autonomy.

Those who count, and for whom we care, we designate as kin. These days we reach for language capacious enough to represent some of the folks with whom we have become kin. Our naming capacities are tested. Our fluency finds us reduced to less than native speakers. Mothers. Fathers. Sisters. Brothers. We know the elasticity we have had to acquire in this vocabulary. Numerous debates and developments show us that we risk illiteracy if we don't arrive at more fluid ways of designating our nearest kin.

The phrase "kissing cousins" allows for some free play in the ways of naming intimates, those who care for us and those we enlist in our care. I have tried to show the extent to which the currency of kinship is no longer blood. Nor is it simply a matter of genes. The medium in and by which we figure our familiars is in flux. As is the language. To achieve a greater fluency in these matters, I have turned to narratives drawn from contemporary fables, fictions, films, and what the French call *fait divers*—stories found in the news of the day, sometimes "ripped from the headlines"—where so many of our adjudications of kinship come to light. Some of the most striking instances of the kinds of stories of kinship that stretch our minds can be found on the front pages.

In the middle of the front page of the *New York Times* on April 4, 2002, surrounded by traumas in the Middle East, was an article that declared "Few Risks Seen to the Children of First Cousins." I am always trolling for material in the public culture, but this was more than passing strange. The question of cousins—worthy to place in the midst of Israeli guards beating their compatriot peace activists and Palestinian kin crying over their flesh and blood, side by side? Just who had been worrying about whether first cousins might incur genetic risks, I wondered, reading on. A color-coded map accompanying this story indicated the legal status of first-cousin marriage across the United States. Juxtaposed as this news event was with visible suffering, I could not help but read its place as one that marked a way of drawing in the boundaries, shutting out difference, some atavistic cautionary tale of why we might do better to "stick to our own kind." Perhaps this was a perverse reading, but surely it was one that tallies with the wish to keep segregated those minimal differences that are tolerable—the kind that enforce endogamy while insisting simultaneously on some degree of exogamy.

This was front-page news? In 2002? That first cousins may intermarry and reproduce without much increased risk of birth or genetic defects? What is it about this scientifically approved long-established practice that needs experts to be quoted and laws to be cited? There, again, we are recalled to the agility, not fragility, of kissing those who are near and already dear. There is a discourse of mixing here that presents as acceptable risk levels what I imagine would be found unacceptable if it were a matter of new research and practices. The map showing states where cousin marriage is legal seemed to beg for cultural analysis, positioned as it was next to the grief of Palestinians (in the week after suicide bombings on Passover and in a Haifa café) that the world was and still is watching.

And what is the dialogue between this demographic in the face of struggles of prophetic proportion between the distant cousins of the sons of Abraham? To me the map suggested, in a matter I found chilling, that we are likely to see a tightening of the circle of cousinship because it is both law and legend in the discourse of "sticking to your own kind." What is news, exactly, in this picture of our world today? Semitic neighbors insist they are not kin but mortal enemies. We know the world is off its axis when the future is murdered—girls and boys killing themselves for a cause. Yes, women, too, will die for an idea, not simply for their children. Again, it is the currency and medium of kinship calculations that intrigues me, calls me to attention.

Kissing cousins: a liminal relationship where we decide how to embrace and contain the others in our care. We choose how we will shield those who are entrusted into our care. How do we come to name our familiars? How do they become family members: a group or body to which we belong and which also exceeds the self?

Kissing cousins: the threshold of destiny and desire. The phrase names a space of permeability. Kinship is the storehouse of cultural narratives of who counts and who cares. Narratives of how care is calculated. In what medium. In what currency.

We speak of getting "under someone's skin." And when we do, we may draw blood. This substance nurtured by sentiment is where I have been drawn to understand some of the world around me. There are countless narratives that make most of us sit up and take notice—stories that reread our pasts and rewrite our futures. I have been moved to ask especially about the narratives that emerge from the struggle between the mixing and the shedding of blood.

Permeability allows us to cross borders. To cross identities. To cross species. Permeability raises questions of intimacy—what passes across membranes may result from the mixing or the shedding of blood. Other mediums are available for the reckoning of kinship, but it is very much the case that blood continues to count when we contemplate who deserves our care and whom we exclude from the regime of the familiar, the familial. These are crucial sites of conflict and reconciliation in the making of social life. "Blood has no nationality" (*Gattaca*).

Janet Carsten, an anthropologist, clarifies the "combination of sentiment, substance, and nurturance" (*Cultures of Relatedness*, 22) that undergirds relatedness or kinship. The substance is the material I have been designating as blood, metaphorically and literally. The sentiment is the affect collected under the sign of care. And nurturance is the labor that an earlier generation of feminist anthropologists called kinwork. Sarah Blaffer Hrdy, with *Mother Nature*, has written the masterwork that ought to make it impossible to ever return to an overidealized discourse of maternal love and instinct. Hrdy teaches her readers to look more closely than we might wish at the trade-offs mothers of all species have always made so that the stew of substance, sentiment, and nurturance continues to sustain life as we know it.

Here are two disparate tales of parental "care" to contemplate: (1) a Kurdish father in Sweden murdered his daughter who threatened to bring shame to the family by becoming involved with a Swedish man, and (2) a mother in the United States murdered two of her three sons in a family with Huntington's chorea. The liquid currency of kinship runs through both these stories. It is a boggy landscape—and dangerous. The dangers to a child who set out to cross new borderlines of care are obvious in the first story; in the second a mother's love adjudicates genetic law. I'd speculate that where blood has been mixed, we dare to learn more about our possible futures. And where blood continues to be shed, there, I believe, we learn more than we wish about our pasts. But the converse is also true, for from the mixing of blood in this genomic age, we are learning about past permeabilities, and in the shedding of blood, we are also losing potential futures.

Bodily fluids: what we spill, carry, pass on to the future. And I have been aiming to figure out some of the oldest and newest ways to reckon whom we care for—and some of the newer ways that genealogies are scripted require humans willing to risk their survival by relinquishing earlier modes of kinship formation. Surely another model is on display in a story from the autumn of 2001, as the United States pursued its new war in Afghanistan. A young couple in Fremont, California, one Pashtun and one Tajik, decided to marry in the midst of our war in their native land, since living in our country allowed them to celebrate a joining of genealogies, enabling them to tell new tales of kissing kin and cousins.

Recall Pilar Cruz and Sam Deeds in *Lone Star*, holding hands and sitting on the hood of his car as the film ends; or recall Leticia and Hank, the black and white romantic couple of *Monster's Ball*—a film that is a study of generations of racism—as they sit close, eating chocolate ice cream together on the back stoop, looking up at the night sky. To survive, kinship narratives require younger generations purged of hatred. They may be without the comforts of conventional kin formations—defathered, unmothered, widowed—but all are survivors of the spasm of murders committed in the name of the law. Laws that name whose blood may not mix risk assuring that it will be shed instead.

Cousins are a capacious category of kin. We may know them or not. We may speak to them or not. We may love them or not. We may visit with them or not. We may count them—as in first, second, third—and calculate their proximity through once or twice removed discourse in our terms. But whatever our connection, we claim them as our kind, as worthy of recognition, as tied to us by some web of alliance and affinity. Everywhere I turned as I worked on this book during a decade that happened also to span a century's turning, I saw scientists, both natural and social, claiming the category of cousin for other primates as well as so-called more distant species. And in Richard Dawkins's *The Ancestor's Tale* we are all global citizens, some kind of cousin at countless genetic genealogical removes. And Jennifer Ackerman's *Chance in the House of Fate* reminds us of our cousinship with yeast, fruit flies, cephalopods even.

And like embryo adoption, a much more recent form of newly made kin, kinship emerges from the ways we have of calculating whom we are likely to consider our own and who is kept outside the circle of empathy. There is a horror of mixing that is reflected in the laws against miscegenation, which, while they may have all been struck down along with other forms of racial segregation, live on in practices and ideologies that coexist beside and outside the laws. First-cousin marriage is invited back into the (news) fold while talk of same-sex marriage has some of the villagers ready to relight their torches, as it were.

Current infertility treatment can lead to the birth of children carrying genes from three individuals. Medical miracle? Teratology for the twenty-first century? Some people will do anything to achieve reproductive success. Let's see if the ark for the new millennium will take more than two of each kind. What difference does it make if that third individual is not accounted for in the rearing of these children? How does this differ from the third (and implicitly fourth) required for adoption to occur? Or the thirds required in lesbian and gay parenting?

For all the critiques of periodization, we leave marks, like Hansel and Gretel, in hopes of recalling where we have come from should we wish to return to our origins. But didn't all of postmodernism instruct us in the problematics of origin-ary tales? But to move into futures of our own making—for those being made for us will also prevail—don't we need to know who or what is going to come along for the ride? Isn't that the folly of time capsules? Isn't that the richly possible moment represented by a new year, decade, century? For we are as knowledgeable as we will ever be, bearing our memories and histories, about that twentieth century—so brilliant and so beastly. And those of us, some the same, but most not, are as avaricious, avid, and voracious for the new knowledges we know are being made around us. It's this passing of the intellectual batons, this handing down and over what we have gathered in this cycle of passing through here on terra firma and cognita.

Consider the following apothegms:

> At the narrow passage there is no brother and there
>   is no friend.
> A friend advises in his interests not yours.
> Know each other as if you were brothers; negotiate
>   deals as if you were strangers.

These can be found on Web sites devoted to Arab proverbs. Yet another demonstration of the plasticity of the cousin category is an Arab proverb of which a version goes like this: "Me and my cousin against the stranger; me and my brother against the cousin." Perhaps it is wise here to recall the dark humor of Cave 73, for this is merely dark. I am drawn to explore such a proverb for all that it lays bare of just how opportunistic and transitory our allegiances can be to those we consider members of our tribe. As I mentioned earlier: when one of the key points of my work here can be summed up proverbially or comically, I fear my aim may be superfluous. Yet I am reassured of the need to look closely at the fact that kinship is established through conflict *and* connection. Notice in this proverb that the stranger is from the outset configured as a source of threat and danger. Notice also that it is the job of men (brothers) to band together first with and then against others of their own kind (cousins), and once having rid themselves of the threat of the outsider, the stranger, to turn on each other, leaving intact only the smallest unit of kin, and we can assume only as long as "he" (the imaginary brother) presents no further threat. In *Fear of Small Numbers*, Arjun Appadurai speaks of the figure of the suicide bomber as sharpening our gaze at the very microscopic level in our understanding of the global; he also notes the anthem as a document of the small number that coheres an identity. But proverbs, like anthems, have cultural specificities. From the German context and history, Kwame Anthony Appiah recovers the following saying as a demonstration of the spirit he calls counter-cosmopolitan: "If you don't want to be my brother, / Then I'll smash your skull in" (*Cosmopolitanism*, 145). Or a French variant that Frans de Waal brings to our attention from Pierre-Joseph Proudhon in the nineteenth century: "If everyone is my brother, I have no brothers" (*Primates and Philosophers*, 163). Consider gated communities as the new form taken by the ancient cave, with its barring of the opening to others. When walls go up in the "holy land," or on the southern border of Texas or California, we cannot afford to ignore the question of who we may kiss and who we may kill. For borders and walls may make some feel secure, but their presence speaks to our sense of threat and danger. I am not suggesting that danger is nonexistent or that rituals of entrance and egress have no purpose or place, but we have seen in the past few years how putting a sense of danger and need for security first has done serious damage to our sense of who is more or less free to move about the world. The precipitous decline of the United States as a place of imagination, where some are drawn to make a "better" life, signals a shift that will be with us for some time to come, making unpredictable futures—as always.

The end of the twentieth century gave way to the growing pains of the twenty-first. Time requests a place in the cast of characters that make up the stories I have been summoning. The new century insists that we keep learning new languages as they construct themselves in the face of new knowledges that require new ethical negotiations. In the slow, heavy pendulum vacillating between differences and samenesses, others and selves, this newer time shifts our attention from politics to ethics so as to more effectively and affectively return us to politics, where empathy and the good rarely are the starting points, and only fitfully the end points, of the kinwork that would keep us all alive. It all seemed so auspicious—the times in which this project first began to accumulate and gather its cellular materials from the convergence of futurities that we welcomed, though not without cautionary voices whispering. Then it seemed an enterprise that asked difficult questions but from a perspective that assumed not progress and amelioration but a momentum of encircling the others with whom we share our time and place.

Processes were under way that engaged deep and difficult questions of who would make commonality with whom. Two decades later, those seem halcyon days. They weren't. Consider a case that names a moment: the 1987 surrogacy suit against Mary Beth Whitehead by the couple whose child she carried in a contract with the husband. This was a family triangle and legalistic vortex we had never entered and where we learned a great deal about children as property in our time.

We've grown into these debates such that as I write, in 2007, an article may casually refer to surrogates as "professional childbearers" and readers somehow accommodate the idea that there are women for whom carrying a child does not "instinctively" equate with the desire to raise that child. What began with the obvious fact of anonymous donor insemination and its remapping the issue of paternity has over recent decades run parallel with forms of destabilizing maternity in ways that are still being struggled over and integrated into our systems of new forms of kinship. Earlier practices of adoption and fostering of children now intersect with technologically ever more refined practices of producing children whose genetic parentage may be completely distinct from that of their social families.

This question of possible futures (mixing) must fully face the past (tribalism) that lives within and among us when we are driven by fear in our very present (the standoff). As I have reiterated, this is a book in praise of mixing. It is only in mixing with and among our strangers that a future even begins to look possible. And it is the multiplicity produced by mixing among strangers that speaks back against the parochialism and provincialism and, yes, tribalism of those who would shun this world of ours in which any metropolitan neighborhood reveals just how possible and already visible is the culture that allows strangers to be something other than threats or dangers. The children of (im)migrants across the world where massive population shifts have occurred can tell stories of how mixing defies an old world order even while the new is in its birth throes.

Finally, I could no longer resist the voices of those who kept asking whether I had seen the television commercials running in the Geico auto insurance campaign featuring cavemen. So, yes, I ventured into YouTube.com for my "research." In just a few minutes, I could watch all these thirty-second ads. I need to retell them for those of you who haven't seen them, because there is one that begs to be included in my bricolage regarding our ancestors, our elders, our imaginary primitive relations if they could speak to us across time. How might we reframe our present and our futures as we make them up going along?

So from the archives of today: a caveman, carrying a laptop and another bag that includes his tennis racket, rolls along a moving sidewalk in an airport. There he passes an ad for Geico car insurance whose tagline is "So easy a caveman could do it." We see in his expression that he is offended by the *Homo sapien*–centrism of the rhetoric. In another ad (there are several, and an increasing number of YouTube mash-ups online, as well as a not very successful sitcom), he and a friend are having a lunch meeting with a corporate representative of the sponsoring company, and they are expressing their consumers' righteous indignation; the man in the suit across the table, apologetically in ignorance and a tone of wishing not to give offense, replies, "We didn't realize you guys were still around." One of the two cavemen suggests that next time, the company might "do a little research." We know this story has "legs" as a passing cause célèbre regarding discrimination and identity politics when we next see him as an interview subject on a television news broadcast. The interviewer asserts that "historically, you guys have struggled to adapt." His caveman interlocutor replies, slowly and showing impatience with further ignorance: "Walking upright? Discovering fire? Inventing the wheel? . . . Sorry we couldn't get that to you sooner."

Now I am curious about the "guys" in question. One ad called "Cavemen at a party" has a friend interrupt the same two characters we've seen before to announce that he and his most recent girlfriend have kissed and made up; the women in question and in the background appear to be a mixed cave–*Homo sapiens* group. Perhaps the most comical of these ads is set in the caveman's therapist's office; when she asks why he's so sensitive about the line "So easy a caveman could do it," he replies, "How would you like it if someone said 'so easy a therapist could do it'?" "That wouldn't make sense," she comes back. "Why not? Because you're smart?" Before she can reply his cell phone rings, and he announces that it's his mother calling and that he's going to put her on "speakerphone."

Is he the son of a cavewoman? An atavistic mutation we ought to be careful to consider as we would any identitarian group? Is this the latest prod to our conscience of racialized discourse as it encounters the arena of political humor? Hairy though he and his friends are, they could be distant and kissing cousins of Jerome from *Gattaca*. Or descendants of the Y-chromosome kinship web of the 2,000-year-old man with whom I began. The past returns to speak, to sing, to shout out to our present, with humor, with warning. The future materializes out of our past, for the better if we've kept listening and learning its new tongues.

WORKS CITED

Ackerman, Jennifer. *Chance in the House of Fate: A Natural History of Heredity*. Boston: Houghton Mifflin, 2001.

Agamben, Giorgio. *Homo Sacer: Sovereign Power and Bare Life*. Translated by Daniel Heller-Roazen. Stanford, Calif.: Stanford University Press, 1998.

——. *The Open: Man and Animal*. Translated by Kevin Attell. Stanford, Calif.: Stanford University Press, 2004.

"All Geico Cavemen Commercial." YouTube, www.youtube.com/watch?v=05JV0Fs_GE8 (accessed February 12, 2007).

Angier, Natalie. "Bonobo Society: Amicable, Amorous, and Run by Females." *New York Times*, April 22, 1997, C4.

——. "Primate Expert Explores Motherhood's Brutal Side." *New York Times*, February 8, 2000, F1.

Apollo 13. Directed by Ron Howard. Universal Pictures, 1995.

Appadurai, Arjun. *Fear of Small Numbers: An Essay on the Geography of Anger*. Durham, N.C.: Duke University Press, 2006.

Appiah, Kwame Anthony. *Cosmopolitanism: Ethics in a World of Strangers*. New York: Norton, 2006.

Associated Press. "At London Zoo, Homo Sapiens Is Just Another Primate Species." *New York Times*, August 28, 2005, sec. 1.

Atwood, Margaret. *The Handmaid's Tale*. Boston: Houghton Mifflin, 1986.

Aureli, Filippo, and Frans de Waal, eds. *Natural Conflict Resolution*. Berkeley: University of California Press, 2000.

Baker, Robin. *Sperm Wars: The Science of Sex*. New York: Basic Books, 1996.

Barber, Richard. *Bestiary: Being an English Version of the Bodleian Library, Oxford, M.S. Bodley 764, with All the Original Miniatures Reproduced in Facsimile*. Woodbridge: Folio Society, 1993.

Bell, Vikki. *Interrogating Incest: Feminism, Foucault, and the Law*. New York: Routledge, 1993.

Benjamin, Walter. *Illuminations*. Translated by Harry Zohn. New York: Schocken Books, 1968.

Blackwood, Evelyn. *Webs of Power: Women, Kin, and Community in a Sumatran Village*. New York: Rowman & Littlefield, 2000.

Blumenfield, Tami. "Na Education in the Face of Modernity." In *Landscapes of Diversity: Indigenous Knowledge, Sustainable Livelihoods, and Resource Governance in Montane Mainland Southeast Asia*, edited by R. Xu Jianchu and Stephen Mikesell, 487–94. Kunming: Yunnan Science and Technology Press, 2002.

Boulle, Pierre. *Planet of the Apes*. Translated by Xan Fielding. New York: Ballantine, 1963.

Brother from Another Planet. Directed by John Sayles. Anarchist's Convention Films, 1984.

Butler, Judith. *Antigone's Claim: Kinship Between Life and Death*. New York: Columbia University Press, 2000.

Carsten, Janet, ed. *Cultures of Relatedness: New Approaches to the Study of Kinship*. Cambridge: Cambridge University Press, 2000.

Cavalieri, Paola, and Peter Singer. "The Great Ape Project." In *Unsanctifying Human Life: Essays on Ethics*, edited by Helga Kuhse, 128–41. Malden, Mass.: Blackwell, 2002.

Chaudhuri, Una, ed. "Animals and Performance." *Drama Review* 51, no. 1 (2007): T193.

Chow, Rey. *Writing Diaspora: Tactics of Intervention in Contemporary Cultural Studies*. Bloomington: Indiana University Press, 1993.

Cook, Robin. *Chromosome 6*. New York: Berkley Books, 1998.

Dawkins, Richard. *The Ancestor's Tale: A Pilgrimage to the Dawn of Evolution*. New York: Houghton Mifflin, 2004.

de Waal, Frans. *The Ape and the Sushi Master: Cultural Reflections of a Primatologist*. New York: Basic Books, 2001.

——. *Bonobo: The Forgotten Ape*. Photographs by Frans Lanting. Berkeley: University of California Press, 1997.

——. *Chimpanzee Politics: Power and Sex Among Apes*. Baltimore: Johns Hopkins University Press, 1982.

——. *My Family Album: Thirty Years of Primate Photography*. Berkeley: University of California Press, 2003.

——. *Our Inner Ape: A Leading Primatologist Explains Why We Are Who We Are*. New York: Riverhead Books, 2005.

——. *Peacemaking Among Primates*. Cambridge, Mass.: Harvard University Press, 1989.

——. *Primates and Philosophers: How Mortality Evolved*. Princeton, N.J.: Princeton University Press, 2006.

——, ed. *Tree of Origin: What Primate Behavior Can Tell Us About Human Social Evolution*. Cambridge, Mass.: Harvard University Press, 2002.

"Declaration on Great Apes." Great Ape Project, www.greatapeproject.org/declaration.php (accessed December 10, 2007).

Douglas, Mary. *Purity and Danger: An Analysis of the Concepts of Pollution and Taboo*. 1966. Reprint, New York: Routledge, 1991.

Eugenides, Jeffrey. "Baster." *New Yorker*, June 17, 1996, 82–89.

——. *Middlesex*. New York: Farrar, Straus and Giroux, 2002.

Feuer, Alan. "Family's Monkey Is Rare, and So Is Its Legal Battle." *New York Times*, July 19, 2000, A1.

Foucault, Michel. *The History of Sexuality*. Vol. 1, *An Introduction*. Translated by Robert Hurley. New York: Pantheon, 1978.

Fouts, Roger. *Next of Kin: What Chimpanzees Have Taught Me About Who We Are*. New York: Morrow, 1997.

Franklin, Sarah. *Technologies of Procreation: Kinship in the Age of Assisted Conception*. New York: Routledge, 1999.

Freud, Sigmund. *Totem and Taboo: Some Points of Agreement Between the Mental Lives of Savages and Neurotics*. Translated by James Strachey. 1950. Reprint, New York: Norton, 1989.

Gattaca. Directed by Andrew Niccol. Columbia Pictures, 1997.

Geertz, Clifford. "The Visit." Review of *A Society Without Fathers or Husbands: The Na of China*, by Cai Hua. *New York Review of Books*, October 18, 2001, 27–30.

Gordimer, Nadine. *The Pickup*. New York: Penguin, 2003.

Grady, Denise. "Few Risks Seen to the Children of First Cousins." *New York Times*, April 4, 2002, A1.

Great Ape Project. www.greatapeproject.org (accessed December 11, 2007).

Haraway, Donna. "A Cyborg Manifesto: Science, Technology, and Socialist-Feminism in the Late Twentieth Century." In *Simians, Cyborgs, and Women: The Reinvention of Nature*, 149–82. New York: Routledge, 1991.

——. *Primate Visions: Gender, Race, and Nature in the World of Modern Science*. New York: Routledge, 1989.

——. *Simians, Cyborgs, and Women: The Reinvention of Nature*. New York: Routledge, 1991.

Haraway, Donna, and Thyrza Nichols Goodeve. *How Like a Leaf: An Interview with Thyrza Nichols Goodeve*. New York: Routledge, 2000.

Hooton, Earnest Albert. *Apes, Men, and Morons*. New York: Books for Libraries Press, 1970.

Hua, Cai. *A Society Without Fathers or Husbands: The Na of China*. Translated by Asti Hustvedt. New York: Zone Books, 2001.

Human Nature. Directed by Michel Gondry. Fine Line Features, 2001.

Jahme, Carole. *Beauty and the Beasts: Woman, Ape, and Evolution.* New York: Soho Press, 2000.

Kafka, Franz. "A Report to an Academy." In *A Country Doctor: Short Stories.* Translated by Siegfried Mortkowitz. Prague: Vitalis, 1998.

Kanō, Takayoshi. *The Last Ape: Pygmy Chimpanzee Behavior and Ecology.* Translated by Evelyn Ono Vineberg. Stanford, Calif.: Stanford University Press, 1992.

Lanier, Jaron. "One Half of a Manifesto." *The Third Culture,* www.edge.org/3rd_culture/lanier/lanier_index.html (accessed December 12, 2007).

Leon, the Pig Farmer. Directed by Vadim Jean and Gary Sinyor. Leon the Pig Farmer Productions, 1993.

Lerner, Max. *The Unfinished Country.* New York: Simon and Schuster, 1959.

Lone Star. Directed by John Sayles. Castle Rock Entertainment, 1996.

Lyall, Sarah. "Lost in Sweden: A Kurdish Daughter Is Sacrificed." *New York Times,* July 23, 2002, A3.

——. "Whites Have Black Twins in In-Vitro Mix-Up." *New York Times,* July 9, 2002, A12.

Mawer, Simon. *Mendel's Dwarf.* 1998. Reprint, New York: Penguin, 1999.

Meloy, Ellen. *The Anthropology of Turquoise.* New York: Vintage, 2002.

Monster's Ball. Directed by Marc Forster. Lee Daniels Entertainment, 2001.

Morrison, Toni. *Song of Solomon.* New York: Vintage International, 2004.

Phillips, Adam. *On Flirtation: Psychoanalytic Essays on the Uncommitted Life.* Cambridge, Mass.: Harvard University Press, 1994.

Piercy, Marge. *Woman on the Edge of Time.* New York: Knopf, 1976.

Planet of the Apes. Directed by Franklin J. Shaffner. APJAC Productions, 1968.

Plotz, David. *The Genius Factory: The Curious History of the Nobel Prize Sperm Bank.* New York: Random House, 2005.

Pusey, Anne E. "Of Genes and Apes: Chimpanzee Social Organization and Reproduction." In *Tree of Origin: What Primate Behavior Can Tell Us About Human Social Evolution*, edited by Frans de Waal, 9–37. Cambridge, Mass.: Harvard University Press, 2002.

Ramachandran, V. S. "Mirror Neurons and Imitation Learning as the Driving Force Behind 'The Great Leap Forward' in Human Evolution." The Third Culture, www.edge.org/3rd_culture/ramachandran/ramachandran_p1.html (accessed March 10, 2007).

Reiner, Carl, and Mel Brooks. *Best of the Two Thousand Year Old Man.* Capitol Records, 1960.

Rey, Margret, and H. A. Rey. *The Complete Adventures of Curious George.* New York: Houghton Mifflin, 2001.

The Right Stuff. Directed by Philip Kaufman. The Ladd Company, 1983.

Rothberg, Michael. Review of *Ethics: An Essay on the Understanding of Evil,* by Alain Badiou, *Criticism* 43, no. 4 (2001): 478–84.

Sapolsky, Robert M. *Monkeyluv: And Other Essays on Our Lives as Animals.* New York: Scribner, 2005.

———. *A Primate's Memoir: A Neuroscientist's Unconventional Life Among the Baboons.* New York: Scribner, 2001.

Savage-Rumbaugh, Sue, and Roger Lewin. *Kanzi: The Ape at the Brink of the Human Mind.* New York: Wiley, 1994.

Schneider, David M. *American Kinship: A Cultural Account.* Chicago: University of Chicago Press, 1980.

Schwartz, John, and Matthew L. Wald. "Echoes of *Challenger*: Shuttle Panel Considers Longstanding Flaws in NASA's Systems." *New York Times,* April 13, 2003, A27.

Self, Will. *Great Apes*. New York: Grove Press, 1997.

Siebert, Charles. "Planet of the Retired Apes." *New York Times Magazine*, July 24, 2005, 26–35.

Simon, Paul. "At the Zoo." *Bookends*. Columbia Records, 1968.

Simone. Directed by Andrew Niccol. New Line Cinema, 2002.

Strathern, Marilyn. *The Gender of the Gift*. New York: Routledge, 1988.

Sunshine State. Directed by John Sayles. Anarchist's Convention Films, 2002.

Sykes, Bryan. *Adam's Curse: The Science That Reveals Our Genetic Destiny*. New York: Norton, 2004.

——. *The Seven Daughters of Eve: The Science That Reveals Our Genetic Ancestry*. New York: Norton, 2001.

The Terminal. Directed by Steven Spielberg. DreamWorks SKG, 2004.

Thompson, Nato, ed. *Becoming Animal: Contemporary Art in the Animal Kingdom*. North Adams, Mass.: MIT Press, 2005.

Trawick, Margaret. *Notes on Love in a Tamil Family*. Berkeley: University of California Press, 1990.

The Truman Show. Directed by Peter Weir. Paramount Pictures, 1998.

2001: A Space Odyssey. Directed by Stanley Kubrick. Metro-Goldwyn-Mayer (MGM), 1968.

"Unexpected Birth at Chimp Haven." Primatology.net, January 17, 2007, http://primatology.net/2007/01/17/unexpected-birth-at-chimp-haven/ (accessed January 8, 2008).

Vaughan, Diane. *The Challenger Launch Decision: Risky Technology, Culture, and Deviance at NASA*. Chicago: University of Chicago Press, 1996.

Weiner, Annette B. *Inalienable Possessions: The Paradox of Keeping-While-Giving*. Berkeley: University of California Press, 1992.

"What's in a Name? About 'Bonobo.'" National Zoo Online, 2001, http://nationalzoo.si.edu/Animals/Primates/Facts/WhatsInA-Name/default.cfm (accessed June 20, 2007).

Williams, Patricia J. *The Alchemy of Race and Rights.* Cambridge, Mass.: Harvard University Press, 1991.

Wrangham, Richard, and Dale Peterson. *Demonic Males: Apes and the Origins of Human Violence.* New York: Mariner Books, 1996.

Yarn, Douglas H. "Law, Love, and Reconciliation: Searching for Natural Conflict Resolution in *Homo Sapiens.*" In *Natural Conflict Resolution*, edited by Filippo Aureli and Frans de Waal, 54–65. Berkeley: University of California Press, 2000.

Zhou, Xiaoli. *The Women's Kingdom.* Rough Cut Video, 2006.

WORKS CONSULTED

Adler, Jerry. "Mind Reading: The New Science of Decision Making. It's Not as Rational as You Think." *Newsweek*, July 5, 2004, 43–47.

Allen, John L., Jr. "Brave, New Biotech World—Human, Animal Mix Raises Ethical Concerns." *National Catholic Reporter Online*, February 13, 2007, www.catholic.org/national/national_story.php?id=23037 (accessed January 15, 2008).

Allen, Teddy. "Baby Chimpanzee Arrives Naturally at Chimp Haven." *Shreveporttimes.com*, January 17, 2007, www.digitaljournal.com/article/93433 (accessed January 15, 2008).

Allison, Dorothy. "First Person: Mama and Mom and Dad and Son." *Bazaar*, March 1998, 114–30.

Alvarez, Lizette. "Spreading Scandinavian Genes, Without Viking Boats." *New York Times*, September 30, 2004, www.scandinaviancryobank.com/spreading-scandinavian-genes.aspx (accessed July 12, 2007).

Angier, Natalie. "Author Offers Theory on Gray Matter of Love." *New York Times*, May 30, 2000, F2.

———. "Cooking, and How It Slew the Beast Within." *New York Times*, May 28, 2002, F1.

———. "In Mandrill Society, Life Is a Girl Thing: The Bachelors Live on the Margins." *New York Times*, May 23, 2000, F1.

——. "Joy in Rwanda: Signing On with the Gorillas." *New York Times*, January 15, 2002, F1.

——. "Not Just Genes: Moving Beyond Nature vs. Nurture." *New York Times*, February 25, 2003, F1.

——. "Sleek, Fast, and Focused: The Cells That Make Dad Dad." *New York Times*, June 12, 2007, F1.

Associated Press. "Ancient Hominid Found in Ethiopia Is Yielding Teeth like the Apes.'" *New York Times*, January 20, 2005, A11.

Badiou, Alain. *Ethics: An Essay on the Understanding of Evil.* Translated by Peter Hallward. London: Verso, 2001.

Bakalar, Nicholas. "Scientists Document Menopause in Captive Gorillas." *New York Times*, January 3, 2006, F8.

Barash, David. "Evolution, Males, and Violence." *Chronicle Review*, May 24, 2002, B7.

Baxter, Ron. *Bestiaries and Their Users in the Middle Ages.* Stroud: Sutton, 1998.

Bellafante, Ginia. "Surrogate Mothers' New Niche: Bearing Babies for Gay Couples." *New York Times*, May 27, 2005, A1.

Belluck, Pam. "From Stem Cell Opponents, an Embryo Crusade." *New York Times*, June 2, 2005, A1.

——. "It's Not So Easy to Adopt an Embryo." *New York Times*, June 12, 2005, www.nytimes.com/2005/06/12/weekinreview/12belluck.html?ex=1184385600&en=836e0952eaaa695e&ei=5070 (accessed July 12, 2007).

Bernard-Donals, Michael. "Ethics After Auschwitz." Review of *Vigilant Memory: Emmanuel Lévinas, the Holocaust, and the Unjust Death*, by Clifton R. Spargo. *Contemporary Literature* 47, no. 1 (2006): 148–52.

Biagini, Enza, and Anna Nozzoli. *Bestiario del novecento.* Rome: Bulzoni Editore, 2001.

Big Love. Episode 15, "Reunion." Directed by Alan Poul. HBO, 2007.

Blake, C. Fred. Review of *A Society Without Fathers or Husbands:*

The Na of China, by Cai Hua. *China Review International* 10, no. 1 (2003): 103–6.

Brettell, Caroline B. "Not That Lineage Stuff: Teaching Kinship into the Twenty-First Century." In *New Directions in Anthropological Kinship*, edited by Linda Stone, 48–70. New York: Rowman & Littlefield, 2001.

Briggs, Laura, and Jodi I. Kelber-Kaye. "'There Is No Unauthorized Breeding in Jurassic Park': Gender and the Uses of Genetics." *National Women's Studies Association Journal* 12, no. 3 (2000): 92–113.

Brody, Jane E. "Battered Women Face Pit Bulls and Cobras." *New York Times*, March 17, 1998, F7.

———. "Genes May Draw Your Road Map but You Still Chart Your Course." *New York Times*, February 25, 2003, F11.

Bruni, Frank. "In an Age of Consent, Defining Abuse by Adults." *New York Times*, November 9, 1997, sec. 3.

Buchan, Nick. "'Human-Brained' Monkeys." *NEWS.com.au*, July 11, 2005, www.news.com.au/story/0,10117,15891104-13762,00. html (accessed January 8, 2008).

Buck-Morss, Susan. *Thinking Past Terror: Islamism and Critical Theory on the Left*. London: Verso, 2003.

Burns, John F. "If It's Civil War, Do We Know It?" *New York Times*, July 24, 2005, sec. 4.

Butler, Judith. *Precarious Life: The Powers of Mourning and Violence*. London: Verso, 2004.

Camporesi, Piero. *The Incorruptible Flesh: Bodily Mutation and Mortification in Religion and Folklore*. Translated by Tania Croft-Murray and Helen Elsom. Cambridge: Cambridge University Press, 1988.

Chang, Kenneth. "Now There Are Many: Robots That Reproduce." *New York Times*, May 17, 2005, F1.

Chen, David W. "See Spot Run, or Is That Spot's Clone?" *New York Times*, December 9, 2000, B1.

Cheney, Dorothy L., and Robert M. Seyfarth. *How Monkeys See the World: Inside the Mind of Another Species*. Chicago: University of Chicago Press, 1990.

"Chimpanzees 'Hunt Using Spears.'" *BBC News Online*, February 22, 2007, http://news.bbc.co.uk/2/hi/science/nature/6387611.stm (accessed February 25, 2007).

Chiu, Lisa Seachrist. *When a Gene Makes You Smell Like Fish: And Other Tales About the Genes in Your Body*. New York: Oxford University Press, 2006.

Churchill, Caryl. *A Number*. New York: Theatre Communications Group, 2003.

Clark, Willene B., and Meradith T. McMunn. *Beasts and Birds of the Middle Ages: The Bestiary and Its Legacy*. Philadelphia: University of Pennsylvania Press, 1989.

Clarke, Angus, and Evelyn Parsons, eds. *Culture, Kinship, and Genes: Towards Cross-Cultural Genetics*. New York: St. Martin's, 1997.

Coetzee, J. M. *Elizabeth Costello: Eight Lessons*. New York: Vintage, 2004.

Collier, Jane Fishburne, and Sylvia Junko Yanagisako, eds. *Gender and Kinship: Essays Toward a Unified Analysis*. Stanford, Calif.: Stanford University Press, 1987.

Cook, Robin. "Decoding Health Insurance." *New York Times*, May 22, 2005, 13.

Corbey, Raymond, and Bert Theunissen, eds. *Ape, Man, Apeman: Changing Views Since 1600*. Leiden: Department of Prehistory, Leiden University, 1995.

Cosh, Colby. "Upfront: Kissing Cousins." *The Report*, May 27, 2002, 3.

Costello, Elvis. "Monkey to Man." *The Delivery Man*. Lost Highway Records, 2004.

Crenson, Matt. "DNA Study Maps Human–Chimp Split." Associated Press, May 17, 2006, www.usatoday.com/tech/science/genetics/2006–05–17-human-chimp-split_x.htm (accessed May 18, 2006).

Critical Art Ensemble. *The Molecular Invasion*. New York: Autonomedia, 2002.

Cronk, Lee. *That Complex Whole: Culture and the Evolution of Human Behavior*. Boulder, Colo.: Westview Press, 1999.

Cusick, Rick, and Warren Chappell. *The Proverbial Bestiary*. Woolwich, Me.: TBW Books, 1982.

D'Agnese, Joseph. "An Embarrassment of Chimpanzees." *Discover*, May 2002, 43–48.

Davis, Colin. "Levinas at 100." *Paragraph* 29, no. 3 (2006): 95–104.

de Gubernatis, Angelo. *Zoological Mythology: The Legends of Animals*. Vol. 2. 1872. Reprint, Detroit: Singing Tree Press, 1968.

de Waal, Frans. *Good Natured: The Origins of Right and Wrong in Humans and Other Animals*. Cambridge, Mass.: Harvard University Press, 1996.

——. "Looking at Flipper, Seeing Ourselves." *New York Times*, October 9, 2006, A17.

——. "Pointing Primates: Sharing Knowledge Without Language." *Chronicle of Higher Education*, January 19, 2001, B7.

Derr, Mark. "Crossbreeding to Save Species and Create New Ones." *New York Times*, July 9, 2002, F3.

DeSalle, Rob, and Michael Yudell. *Welcome to the Genome: A User's Guide to the Genetic Past, Present, and Future*. New York: Wiley, 2005.

Diamond, Jared. *The Third Chimpanzee: The Evolution and Future of the Human Animal*. New York: HarperCollins, 1992.

Díaz, Nancy Gray. *The Radical Self: Metamorphosis to Animal Form in Modern Latin American Narrative*. Columbia: University of Missouri Press, 1988.

Dickinson, Peter. *Eva*. New York: Delacorte Press, 1988.

Dowd, Maureen. "The Baby Bust." *New York Times*, April 10, 2002, A27.

——. "Saving the Orangutan, Preserving Paradise." *New York Times*, March 21, 2000, F3.

——. "What Rough Beasts?" *New York Times*, May 7, 2005, A15.

——. "Y? DNA! Q.E.D." *New York Times*, April 17, 2002, A23.

Dreifus, Claudia. "A Conversation with Cynthia Breazeal: A Passion to Build a Better Robot, One with Social Skills and a Smile." *New York Times*, June 10, 2003, F3.

——. "Scientist at Work/Biruté Galdikas: Saving the Orangutan, Preserving Paradise." *New York Times*, March 23, 2000, F3.

Dubner, Stephen J., and Steven D. Levitt. "Monkey Business: Can Capuchins Understand Money?" *New York Times Magazine*, June 5, 2005, 30–32.

Dugger, Celia W. "Beware Monkey-Man, Scourge of the Gullible." *New York Times*, May 19, 2001, A4.

Eakin, Emily. "Before the Word, Perhaps the Wink?" *New York Times*, May 18, 2002, B7.

Edwards, Jeanette. *Born and Bred: Idioms of Kinship and New Reproductive Technologies in England*. New York: Oxford University Press, 2000.

Eilberg-Schwartz, Howard. *The Savage in Judaism: An Anthropology of Israelite Religion and Ancient Judaism*. Bloomington: Indiana University Press, 1990.

Enders, Eric. "Forget the Alamo: The Convergence of Border Cultures in John Sayles' *Lone Star*." *EricEnders.com*, 1999, www.ericenders.com/lonestar.htm (accessed April 2, 2003).

Engels, Frederick. *The Part Played by Labour in the Transition from Ape to Man*. Translated by Clemens Dutt. Moscow: Progress Publishers, 1934.

"Fear Itself." *New York Times*, May 20, 2001, A2.

Ferris, Timothy. "At Dawn, the *Columbia*." *New York Times*, February 3, 2003, A25.

Finkler, Kaja. *Experiencing the New Genetics: Family and Kinship on the Medical Frontier.* Philadelphia: University of Pennsylvania Press, 2000.

Fisher, Ian. "Account of Punjab Rape Tells of a Brutal Society." *New York Times*, July 17, 2002, A3.

——. "Seeing No Justice, a Rape Victim Chooses Death." *New York Times*, July 28, 2002, http://query.nytimes.com/gst/fullpage.html?sec=health&res=9906E3DE173BF93BA15754C0A96 49C8B63 (accessed July 30, 2002).

Flanagan, Dennis, ed. *Twentieth-Century Bestiary.* 1948. Reprint, New York: Simon and Schuster, 1955.

Fleming, Anne Taylor. "Why I Can't Use Someone Else's Eggs." *Newsweek,* April 12, 1999, 12.

Fossey, Dian. *Gorillas in the Mist.* Boston: Houghton Mifflin, 1983.

Foster, Douglas. "Are You in Anthropodenial?" Review of *The Ape and the Sushi Master,* by Frans de Waal. *New York Times Book Review,* April 8, 2001, 30.

Fountain, Henry. "Splish-Splash, Gorillas Are Taking a Bath and Sending a Message." *New York Times,* August 7, 2001, F6.

Fournival, Richard de. *Master Richard's Bestiary of Love and Response.* Translated by Jeanette Beer. Berkeley: University of California Press, 1986.

Fox, Robin. *The Red Lamp of Incest: An Enquiry into the Origins of Mind and Society.* Notre Dame, Ind.: University of Notre Dame Press, 1980.

Francoeur, Robert T. *Utopian Motherhood: New Trends in Human Reproduction.* New York: Perpetua Books, 1974.

Garreau, Joel. *Radical Evolution: The Promise and Peril of Enhancing Our Minds, Our Bodies—and What It Means to Be Human.* New York: Doubleday, 2005.

Glover, Jonathan, et al. *Ethics of New Reproductive Technologies: The Glover Report to the European Commission.* DeKalb: Northern Illinois University Press, 1989.

Goodall, Jane. *The Chimpanzees of Gombe: Patterns of Behavior.* Cambridge, Mass.: Belknap Press, 1986.

Goode, Erica. "Humans Found Closer Yet to Animals." *New York Times*, June 25, 1999, A19.

Gorman, James. "It's Not Just Apes: Fruit Flies Are Our Cousins, Too." *New York Times*, August 22, 2006, F4.

Gould, Stephen Jay. "The Human Difference." *New York Times*, July 2, 1999, A17.

Grady, Denise. "Photograph 51." *New York Times*, February 25, 2003, F2.

——. "Woman Has Child After Receiving Twin's Ovarian Tissue." *New York Times*, June 8, 2005, A12.

Grand, Sue. *The Reproduction of Evil: A Clinical and Cultural Perspective.* Hillsdale: Analytic Press, 2000.

Grandin, Temple, and Catherine Johnson. *Animals in Translation: Using the Mysteries of Autism to Decode Animal Behavior.* Orlando, Fla.: Harcourt, 2005.

Grant, Annette. "America the Beautifully Absurd." *New York Times*, May 22, 2005, sec. 2.

Gray, John. *Straw Dogs: Thoughts on Humans and Other Animals.* London: Granta Books, 2002.

Gross, Jane. "Comfort and Refuge in Black Hamptons Enclave." *New York Times*, July 16, 2002, B1.

Hamilton, Jon. "A Voluble Visit with Two Talking Apes." *NPR.com*, July 8, 2006, www.npr.org/templates/story/story.php?storyId=5503685 (accessed July 8, 2006).

Hanscombe, Gillian E., and Jackie Forster. *Rocking the Cradle: Lesbian Mothers: A Challenge in Family Living.* Boston: Alyson, 1982.

Harmon, Amy. "Blacks Pin Hope on DNA to Fill Slavery's Gaps in Family Trees." *New York Times*, July 25, 2005, A1.

——. "In New Tests for Fetal Defects, Agonizing Choices for Parents." *New York Times*, June 20, 2004, A1.

———. "Sperm Donor Father Ends His Anonymity." *New York Times*, February 14, 2007, A18.

———. "Twist and Shout! The Double Helix Replicates Itself in Popular Culture." *New York Times*, February 25, 2003, F10.

Harrison, Sophie. "The Wrinkle Cure." Review of *Dorian*, by Will Self. *New York Times Book Review*, January 5, 2003, 6.

Hayles, Katherine N. *The Cosmic Web: Scientific Field Models and Literary Strategies in the Twentieth Century*. Ithaca, N.Y.: Cornell University Press, 1984.

Helms, Mary W. *Access to Origins: Affines, Ancestors, and Aristocrats*. Austin: University of Texas Press, 1998.

Henig, Robin Marantz. "Deciding When Science Has Gone Astray." *New York Times*, February 25, 2003, F5.

Holden, Stephen. "A Family Born of an Unusual Collaboration." Review of *Parental Instinct*, dir. Murray Nossel. *New York Times*, June 8, 2005, E5.

Holloway, Marguerite. "Profile: Ruth Hubbard." *Scientific American*, June 1995, 49–50.

Hooton, Earnest Albert. *Up from the Ape*. 1931. Reprint, New York: Macmillan, 1946.

———. *Why Men Behave Like Apes and Vice Versa: Body and Behavior*. Princeton, N.J.: Princeton University Press, 1940.

Hrdy, Sarah Blaffer. *Mother Nature: Maternal Instincts and How They Shape the Human Species*. New York: Ballantine, 1999.

Hunter, Brad. "Apes Instrumental to Rocker Gabriel's New Band." *New York Post*, June 18, 2001, 19.

Johnson, Harriet McBryde. "Unspeakable Conversations or How I Spent One Day as a Token Cripple at Princeton University." *New York Times Magazine*, February 16, 2003, 50–79.

Jones, Heather, and Maggie Kirkman. "Embedded in the Sperm Wars." *On Line Opinion*, May 26, 2005, www.onlineopinion.com.au/view.asp?article=3493 (accessed June 10, 2005).

Köhler, Wolfgang. *The Mentality of Apes*. Translated by Ella Winter. New York: Liveright, 1973.

Kolata, Gina. "Cheating, or an Early Mingling of the Blood?" *New York Times*, May 10, 2005, www.nytimes.com/2005/05/10/health/10bloo.html?ex=1184385600&en=3493b8696c6dc565&ei=5070 (accessed July 12, 2007).

——. "Genetic Revolution: How Much, How Fast?" *New York Times*, February 25, 2003, F6.

——. "Name Games and the Science of Life." *New York Times*, May 29, 2005, 12.

Krist, Gary. "The Slippery Slope." Review of *The Fall*, by Simon Mawer. *New York Times Book Review*, January 5, 2003, 7.

Kuhlmann, Hannah. "Escaping Abjection: *Gattaca*'s Narratives of Discrimination." *University of Minnesota*, wysiwyg://12/http://www.tc.umn.edu/~matri001/wost3190/Kuhlmann3.html (accessed July 11, 2007).

Le Douarin, Nicole, and Anne McLaren, eds. *Chimeras in Developmental Biology*. New York: Academic Press, 1984.

Lévinas, Emmanuel. *Alterity and Transcendence*. Translated by Michael B. Smith. New York: Columbia University Press, 1999.

——. *Proper Names*. Translated by Michael B. Smith. Stanford, Calif.: Stanford University Press, 1996.

Lewin, Tamar. "Two Views of Growing Up When the Faces Don't Match." *New York Times*, October 27, 1998, www.nytimes.com/library/national/102798adoption-sb.html (accessed November 5, 1998).

Lyall, Sarah. "British Judge Rules Sperm Donor Is Legal Father in Mix-Up Case." *New York Times*, February 27, 2003, A5.

Madigan, Nick. "After Fleeing Polygamist Community, an Opportunity for Influence." *New York Times*, June 29, 2005, A16.

Madoff, Steven Henry. "The Wonders of Genetics Breed a New Art." *New York Times*, May 26, 2002, B1.

Majendie, Paul. "Did Humans Descend from 33 Daughters of

Eve?" *World Wide Religious News*, May 29, 2001, www.wwrn. org/article.php?idd=14344&sec=76&con=55 (accessed August 20, 2002).

Mansnerus, Laura. "The Baby Bazaar: How Bundles of Joy Not for Sale Are Sold." *New York Times*, October 26, 1998, www. nytimes.com/library/national/102698adoption.html (accessed November 5, 1998).

Marinoni, Augusto. *Leonardo: Bestiario e favole*. Milan: TEA, 1988.

Marks, Jonathan. *What It Means to Be 98% Chimpanzee: Apes, People, and Their Genes*. Berkeley: University of California Press, 2002.

Maslin, Janet. "A Bizarre Tale of the Rise and Fall of an Elitist Sperm Bank." Review of *The Genius Factory*, by David Plotz. *New York Times*, June 2, 2005, E10.

——. "A Cross-Cultural Marriage, a Notorious Brother-in-Law." Review of *Inside the Kingdom: My Life in Saudi Arabia*, by Carmen bin Laden. *New York Times*, July 1, 2004, E10.

Mattes, Jane. *Single Motherhood by Choice: A Guidebook for Single Women Who Are Considering or Have Chosen Motherhood*. New York: Times Books, 1994.

Mawer, Simon. *Gregor Mendel: Planting the Seeds of Genetics*. New York: Abrams, 2006.

Mayell, Hillary. "Hobbit-like Human Ancestor Found in Asia." *NationalGeographic.com*, 2004, http://news.nationalgeographic.com/news/2004/10/1027_041027_homo_floresiensis.html (accessed November 1, 2004).

Maynes, Mary Jo, Ann Waltner, Birgitte Soland, and Ulrike Strasser, eds. *Gender, Kinship, Power: A Comparative and Interdisciplinary History*. New York: Routledge, 1996.

McCulloch, Florence. *Mediaeval Latin and French Bestiaries*. Chapel Hill: University of North Carolina Press, 1960.

McNeil, Maureen, Ian Varcoe, and Steven Yearley, eds. *The New Reproductive Technologies*. New York: St. Martin's, 1990.

McQuillan, Alice, and Bob Kappstatter. "He Goes Ape at Zoo: Strips, Romps in Gorilla Preserve in Bronx." *Daily News*, August 9, 2001, 5.

Michael, Max, III, W. Thomas Boyce, and Allen J. Wilcox. *Biomedical Bestiary: An Epidemiologic Guide to Flaws and Fallacies in the Medical Literature*. Boston: Little, Brown, 1984.

Miller, Laura. "Virtue's Hack: John Sayles Makes Movies with All the Right Messages—and No Surprises, Madness or Life." *Salon*, www.salon.com/weekly/movies960729.html (accessed April 18, 1996).

Mirapaul, Matthew. "Digital Artworks That Play Against Expectations." *New York Times*, September 30, 2002, E2.

Mitchell, W. J. T. "Seeing Disability." *Public Culture* 13, no. 3 (2001): 391–97.

——. "The Work of Art in the Age of Biocybernetic Reproduction." *Modernism/Modernity* 10, no. 3 (2003): 481–500.

"Monkey Business." *ABCNEWS.com*, 2001, http://my.abcnews.go.com (accessed May 16, 2001).

Moorman, Margaret. *Waiting to Forget: A Memoir*. New York: Norton, 1996.

Morris, Bob. "We Are Family." *New York Times*, May 22, 2005, sec. 9.

Morris, Brian. *Animals and Ancestors: An Ethnography*. Oxford: Berg, 2000.

——. *The Power of Animals: An Ethnography*. Oxford: Berg, 1998.

Morris, Ramona, and Desmond Morris. *Men and Apes*. New York: Bantam, 1968.

Moss, Cynthia. *Elephant Memories: Thirteen Years in the Life of an Elephant Family*. New York: Morrow, 1988.

Mydans, Seth. "He's Not Hairy, He's My Brother." *New York Times*, August 12, 2001, sec. 4.

Needham, Rodney. *Remarks and Inventions: Skeptical Essays About Kinship*. London: Tavistock, 1974.

Neruda, Pablo. *Bestiario*. Translated by Elsa Neuberger. New York: Harcourt Brace Jovanovich 1965.

"99.5 Percent Like a Neanderthal." *New York Times*, November 18, 2006, A16.

Norris, Michele. "A Chimpanzee Mystery: Who's Your Daddy?" *NPR.com*, January 19, 2007, www.npr.org/templates/story/story.php?storyId=6921863 (accessed January 25, 2007).

Ore, Mary. "Anatomy as Art, Unsettling but Drawing Crowds." *New York Times*, July 9, 2002, E2.

O'Sullivan, Catherine. "A Look at the Brighter Side of Mixed-Species Creations." *Tucson Weekly*, April 21, 2005, 8.

Overbye, Dennis. "For the History of Science, the First Draft Is Often Late." *New York Times*, February 25, 2003, F4.

——. "The Paper and the Papers." *New York Times*, February 25, 2003, F4.

Oxford Ancestors, 2000–2004, http://oxfordancestors.com. (accessed December 18, 2001).

Parrington, John. "The Human Touch." *Socialist Review* 204 (1997): 4.

"Parrot Calls Out Police in Emergency." *Reuters.com*, May 17, 2001, www.blueray.com/wordsworth/news/police.html (accessed January 15, 2008).

Patchett, Ann. "Kissing Cousins" *New York Times Magazine*, April 28, 2002, 21–22.

Peletz, Michael G. "Kinship Studies in Late Twentieth-Century Anthropology." *Annual Review of Anthropology* 24 (1995): 343–72.

Pert, Candace B. *Molecules of Emotion: The Science Behind Mind–Body Medicine*. New York: Simon and Schuster, 1997.

Petersen, Hanne, ed. *Love and Law in Europe*. Brookfield: Dartmouth, 1998.

Peterson, Dale. *The Deluge and the Ark: A Journey into Primate Worlds*. Boston: Houghton Mifflin, 1989.

Peterson, Dale, and Jane Goodall. *Visions of Caliban: On Chimpanzees and People.* New York: Houghton Mifflin, 1993.

Pinker, Steven. "Sniffing Out the Gay Gene." *New York Times,* May 17, 2005, A21.

Plooij, Frans X. *The Behavioral Development of Free-Living Chimpanzee Babies and Infants.* Norwood: Ablex, 1984.

Plotz, David. "Who's Your Daddy?" *New York Times,* May 19, 2005, A27.

Podwal, Mark. *A Jewish Bestiary: A Book of Fabulous Creatures Drawn from Hebraic Legend and Lore.* Philadelphia: Jewish Publication Society of America, 1984.

"Police Focus on Mysterious 'Monkey Man.'" *CNN.com,* 2001, http://cnn.worldnews.com (accessed May 17, 2001).

Pollack, Andrew. "Genome Pioneer Will Start Center of His Own." *New York Times,* http://query.nytimes.com/gst/fullpage.html?sec=health&res=9E0CEFDA163DF936A2575BC0A9649C8B63 (accessed August 15, 2002).

——. "How the Arms of the Helixes Are Poised to Serve." *New York Times,* February 25, 2003, F5.

Pollack, Sandra, and Jeanne Vaughn, eds. *Politics of the Heart: A Lesbian Parenting Anthology.* Ithaca, N.Y.: Firebrand Books, 1987.

Poole, Joyce. *Coming of Age with Elephants: A Memoir.* New York: Hyperion, 1996.

Povoledo, Elisabetta. "Updating an Old Way to Leave the Baby on the Doorstep." *New York Times,* February 28, 2007, A4.

Prose, Francine. "Lithuanians and Letts Do It." Review of *Marriage, a History,* by Stephanie Coontz. *New York Times Book Review,* June 12, 2005, 20.

Rapp, Rayna, and Faye D. Ginsburg. "Enabling Disability: Rewriting Kinship, Reimagining Citizenship." *Public Culture* 13, no. 3 (2001): 533–56.

Rennie, John, ed. "New Look at Human Evolution." Special issue, *Scientific American*, August 25, 2003.

"Revisiting a Law That Was Ahead of Its Time." *TheAge.com*, June 6, 2005, www.theage.com (accessed July 10, 2005).

Ridley, Matt. *Genome: The Autobiography of a Species in 23 Chapters*. New York: Perennial, 1999.

Robertson, John A. *Children of Choice: Freedom and the New Reproductive Technologies*. Princeton, N.J.: Princeton University Press, 1994.

Rogers, Connie. "Revealing Behavior in 'Orangutan Heaven and Human Hell.'" *New York Times*, November 15, 2005, F2.

Rose, Steven. *The Making of Memory: From Molecules to Mind*. New York: Anchor-Doubleday, 1993.

Rosenfield Lafo, Rachel, Nick Capasso, and Dina Deitsch, eds. *Going Ape: Confronting Animals in Contemporary Art*. Lincoln, Mass.: DeCordova Museum and Sculpture Park, 2006. [Published in conjunction with the exhibition "Going Ape: Confronting Animals in Contemporary Art," shown at the DeCordova Museum and Sculpture Park, September 2, 2005–January 7, 2007]

Sachs, Susan. "Clinics' Pitch to Indian Émigrés: It's a Boy." *New York Times*, http://query.nytimes.com/gst/fullpage.html?sec =health&res=9904E3DF173EF936A2575BC0A9679C8B63 (accessed August 15, 2001).

Saffron, Lisa. *Challenging Conceptions: Pregnancy and Parenting Beyond the Traditional Family*. New York: Cassell, 1994.

Safire, William. "Chimera: Rough Beast Slouches Toward Science." *New York Times Magazine*, May 22, 2005, 22.

Savage-Rumbaugh, Sue, Stuart G. Shanker, and Talbot J. Taylor. *Apes, Language, and the Human Mind*. New York: Oxford University Press, 1998.

Schwartz, John. "At One Point, 'I Was Deathly Afraid,' New Space Visitor Admits." *New York Times*, June 23, 2004, A17.

———. "She's Studying. He's Playing." *New York Times*, June 14, 2005, F1.

Scott, A. O. "A Future More Nasty, Because It's So Near." Review of *Code 46*, dir. Michael Winterbottom. *New York Times*, August 6, 2004, E13.

Seibert, Charles. "The Animal Self." *New York Times Magazine*, January 22, 2006, 48–61.

———. "An Elephant Crackup?" *New York Times Magazine*, October 8, 2006, 42–55.

Seigel, Jessica. "Secrets of the Bonobo Sisterhood." *Ms.*, spring 2005, 44–49.

Sengupta, Somini. "The Gentlest of Beasts, Making Love, Ravaged by War." *New York Times*, May 3, 2004, A4.

Shea, Christopher. "Don't Talk to the Humans: The Crackdown on Social Science Research." *Linguafranca*, September 2000, 27–34.

Shore, Cris. "Virgin Births and Sterile Debates." *Current Anthropology* 33, no. 3 (1992): 295–314.

Shteir, Rachel. *Striptease: The Untold History of the Girlie Show*. Oxford: Oxford University Press, 2004.

Simms, Paul. "Talking Chimp Gives His First Press Conference." *New Yorker*, June 6, 2005, 44–45.

Small, Meredith F. *Female Choices: Sexual Behavior and Female Primates*. Ithaca, N.Y.: Cornell University Press, 1993.

Smith, Dinitia. "A Critic Takes On the Logic of Female Orgasm." *New York Times*, May 17, 2005, F1.

Smothers, Ronald. "Court Lets Two Gay Men Jointly Adopt Child." *New York Times*, October 23, 1997, B5.

Sontag, Susan. *Regarding the Pain of Others*. New York: Picador, 2003.

Spallone, Patricia, and Deborah Lynn Steinberg, eds. *Made to Order: The Myth of Reproductive and Genetic Progress*. Oxford: Pergamon Press, 1987.

Spina, Michele. *Sleep: A Utopian Bestiary.* Translated by Ann Colcord and Hugh Shankland. Vale: Guernsey Press, 2001.

Spooner, Phillip. "Zebra Gives Birth to Foal Sired by Donkey." Associated Press, 2005, www.firstcoastnews.com/news/strange/news-article.aspx?storyid=36339 (accessed April 29, 2005).

Squiers, Carol. *The Art of Science.* New York: International Center of Photography, 2004.

Stacey, Judith. *Brave New Families: Stories of Domestic Upheaval in Late Twentieth Century America.* New York: Basic Books, 1990.

Stevenson, James. "Sherry Britton." *New York Times,* November 25, 2006, A15.

Stewart, Kathleen. *A Space on the Side of the Road: Cultural Poetics in an 'Other' America.* Princeton, N.J.: Princeton University Press, 1996.

Strier, Karen B. *Faces in the Forest: The Endangered Muriqui Monkeys of Brazil.* New York: Oxford University Press, 1992.

Sunstein, Cass R. "The Chimps' Day in Court." Review of *Rattling the Cage: Toward Legal Rights for Animals,* by Steven M. Wise. *New York Times Book Review,* February 20, 2000, www.nytimes.com/books/00/02/20/reviews/000220.20sunstet.html (accessed March 25, 2000).

Texas Military Forces Museum, www.texasmilitaryforcesmuseum.org (accessed May 25, 2006).

"Texas Revolution." *The Alamo Online,* www.thealamo.org/main.html (accessed May 25, 2006).

"Through Saturn's Rings." *New York Times,* July 1, 2004, A20.

Tierney, John. "'Get Out, You Damned One.'" *New York Times,* July 2, 2005, A15.

Tomasello, Michael. "For Human Eyes Only." *New York Times,* January 13, 2007, A15.

Tomkins, Calvin. "Flying into the Light." *New Yorker,* January 13, 2003, 62–71.

Trautmann, Thomas R. *Lewis Henry Morgan and the Invention of Kinship*. Berkeley: University of California Press, 1988.

Tremayne, Soraya, ed. *Managing Reproductive Life: Cross-Cultural Themes in Fertility and Sexuality*. New York: Berghahn Books, 2001.

Trencher, Susan R. *Mirrored Images: American Anthropology and American Culture, 1960–1980*. Westport: Bergin & Garvey, 2000.

Vazquez, Lourdes. *Bestiary*. Translated by Rosa Alcala. Tempe, Ariz.: Bilingual Press, 2004.

Verner, Lisa. *The Epistemology of the Monstrous in the Middle Ages*. New York: Routledge, 2005.

Vigran, Anna. "Living with Nyota the Bonobo." *NPR.com*, July 8, 2006, www.npr.org/templates/story/story.php?storyId=5541593 (accessed July 8, 2006).

Villarosa, Linda. "Once-Invisible Sperm Donors Get to Meet the Family." *New York Times*, May 21, 2002, F5.

Wade, Nicholas. "Chimeras on the Horizon, but Don't Expect Centaurs." *New York Times*, May 3, 2005, F1.

——. "Chimpanzees with Culture." *New York Times*, June 27, 1999, A16.

——. "Covert Ops Enter the Genomic Era." *New York Times*, April 20, 2003, sec. 4.

——. "Dating of Australian Remains Backs Theory of Early Migration of Humans." *New York Times*, February 19, 2003, A6.

——. "DNA Twists to the Right. What if It Didn't?" *New York Times*, February 25, 2003, F1.

——. "Double Helix Leaps from Lab into Real Life." *New York Times*, February 25, 2003, F1.

——. "Early Voices: The Leap to Language." *New York Times*, July 15, 2003, F1.

——. "Equality Between the Sexes: Neanderthal Women Joined Men in the Hunt." *New York Times*, December 5, 2006, F3.

———. "Ethicists Offer Advice for Testing Human Brain Cells in Primates." *New York Times*, July 15, 2005, www.nytimes.com/2005/07/15/science/15stem.html?ex=1184385600&en=b40a25a1806aa170&ei=5070 (accessed July 15, 2007).

———. "Evolution of Gene Related to Brain's Growth Is Detailed." *New York Times*, January 14, 2004, A13.

———. "First Couple." *New York Times*, February 25, 2003, F3.

———. "Human or Chimp? 50 Genes Are the Key." *New York Times*, October 20, 1998, www.nytimes.com/library/national/science/10298scimonkey.html (accessed October 23, 1998).

———. "Language Gene Is Traced to Emergence of Humans." *New York Times*, August 15, 2002, http://query.nytimes.com/gst/fullpage.html?sec=health&res=9C07E4DD163DF936A2575BC0A9649C8B63 (accessed August 15. 2002).

———. "New DNA Test Yields Neanderthal Clues, Showing Genome Nearly Identical to Humans.'" *New York Times*, November 16, 2006, A13.

———. "Scientist Finds the Beginnings of Morality in Primate Behavior." *New York Times*, March 20, 2007, F3.

———. "Startling Scientists, a Plant Is Found to Fix Its Own Mutated Gene." *New York Times*, March 23, 2005, A13.

———. "Study Suggests Monkeys Have Ability to Think." *New York Times*, October 23, 1998, www.nytimes.com/library/national/science/10398sci-monkey.htm (accessed October 23, 1998).

———. "Two Splits Between Human and Chimp Lines Suggested." *New York Times*, May 18, 2006, www.npr.org/templates/story/story.php?storyId=5541593 (accessed July 12, 2007).

Walsh, Eileen. Review of *A Society Without Fathers or Husbands: The Na of China*, by Cai Hua. *American Ethnologist* 29, no. 4 (2002): 1043–45.

Walzer, Michael. *Spheres of Justice: A Defense of Pluralism and Equality*. New York: Basic Books, 1983.

Warner, Marina. *Six Myths of Our Time: Little Angels, Little Mon-*

sters, Beautiful Beasts, and More. New York: Random House, 1994.

West, Dennis, and Joan M. West. "Borders and Boundaries: An Interview with John Sayles." *Cineaste* 22, no. 3 (1996): 14.

"Whose [*sic*] Your Daddy? Chimp Haven Wonders." *Boston. com*, January 16, 2007, www.boston.com/news/odd/articles/2007/01/16/whose_your_daddy_chimp_haven_wonders/ (accessed January 20, 2007).

Wilbur, Richard. *A Bestiary*. New York: Pantheon, 1955.

Wilford, John Noble. "As the Veil on Titan Is Lifted, 'What We See Is Very Alien.'" *New York Times*, November 2, 2004, F2.

——. "A Big Debate on Little People: Ancient Species or Modern Dwarfs." *New York Times*, October 12, 2005, A11.

——. "First Close-up Photos of Saturn's Giant Moon." *New York Times*, July 3, 2004, A13.

——. "Fossil Found of a Big Bird That Kermit Wouldn't Like." *New York Times*, October 31, 2006, F3.

——. "Fossils May Be Earliest Human Link." *New York Times*, July 12, 2001, A12.

——. "New Clues to History of Male and Female." *New York Times*, August 26, 1997, C1.

——. "7-Year, 2-Million-Mile Flight Brings Craft in Saturn's Grip." *New York Times*, June 30, 2004, A19.

——. "Spacecraft Becomes First to Enter Orbit of Saturn." *New York Times*, July 1, 2004, A18.

Wilson, Peter J. *Man, the Promising Primate: The Conditions of Human Evolution*. New Haven, Conn.: Yale University Press, 1983.

Wines, Michael. "Cameroon vs. South Africa in the Battle of the Gorillas." *New York Times*, May 10, 2005, A3.

Wood, Alan B. "Genes and Human Potential: Bergsonian Readings of *Gattaca* and the Human Genome." *Theory and Event* 7, no.1 (2003), http://muse.jhu.edu/journals/theory_and_event/voo7/7.1wood.html (accessed November 11, 2003).

Wright, Robert. "Planet of the Apes." *New York Times*, April 28, 2007, A17.

Wynne, Clive D. L. "Kissing Cousins." *New York Times*, December 12, 2005, A27.

Yamamoto, Dorothy. *The Boundaries of the Human in Medieval English Literature*. New York: Oxford University Press, 2000.

Yardley, Jim. "Because of 9/11, a Uniting River Now Divides." *New York Times*, August 1, 2002, A1.

Yoon, Carol Kaesuk. "In This Battle of the Sexes, the Females Win." *New York Times*, December 22, 1998, www.nytimes.com/library/national/science/122298sci-jacana.html (accessed December 31, 1998).

Zanganeh, Lila Azam. "Out of Africa." *New York Times Book Review*, June 12, 2005, 39.

Zellnik, David. *Serendib* [play]. 2007.

Zhao, Yilu. "Living in 2 Worlds, Old and New: Foreign-Born Adoptees Explore Their Cultural Roots." *New York Times*, April 9, 2002, B1.

Zigman, Laura. *Animal Husbandry*. New York: Dial Press, 1998.

Zimmer, Carl. "In the Marmoset Family, Things Really Do Appear to Be All Relative." *New York Times*, March 27, 2007, F4.

INDEX